TWENTY FIRST CENTURY SCIENCE

GCSE Additional Science
Higher

Nuffield Curriculum Centre

THE UNIVERSITY of York

OXFORD

The exercises in this Workbook cover the OCR requirements for each module. If you do them during the course, your completed Workbook will help you revise for exams.

Project Directors
Jenifer Burden
John Holman
Andrew Hunt
Robin Millar

Project Officers
Peter Campbell
Angela Hall
John Lazonby
Peter Nicolson

Course Editors
Jenifer Burden
Peter Campbell
Angela Hall
Andrew Hunt
Robin Millar

Authors
Jenifer Burden
Peter Campbell
Andrew Hunt
Caroline Shearer
Charles Tracy

Contents

B4	Homeostasis	3
C4	Chemical patterns	18
P4	Explaining motion	33
B5	Growth and development	49
C5	Chemicals of the natural environment	64
P5	Electric circuits	79
B6	Brain and mind	95
C6	Chemical synthesis	110
P6	The wave model of radiation	125

WORKBOOK

Great Clarendon Street, Oxford OX2 6DP

Oxford University Press is a department of the University of Oxford.
It furthers the University's objective of excellence in research, scholarship,
and education by publishing worldwide in

Oxford   New York

Auckland   Cape Town   Dar es Salaam   Hong Kong   Karachi
Kuala Lumpur   Madrid   Melbourne   Mexico City   Nairobi
New Delhi   Shanghai   Taipei   Toronto

With offices in

Argentina   Austria   Brazil   Chile   Czech Republic   France   Greece
Guatemala   Hungary   Italy   Japan   Poland   Portugal   Singapore
South Korea   Switzerland   Thailand   Turkey   Ukraine   Vietnam

© University of York on behalf of UYSEG and the Nuffield Foundation 2006

The moral rights of the authors have been asserted

Database right Oxford University Press (maker)

First published 2007

All rights reserved. No part of this publication may be reproduced,
stored in a retrieval system, or transmitted, in any form or by any means,
without the prior permission in writing of Oxford University Press,
or as expressly permitted by law, or under terms agreed with the
appropriate
reprographics rights organization. Enquiries concerning reproduction
outside the scope of the above should be sent to the Rights Department,
Oxford University Press, at the address above

You must not circulate this book in any other binding or cover
and you must impose this same condition on any acquirer

British Library Cataloguing in Publication Data

Data available

ISBN: 978019915293

10 9 8 7 6 5 4 3 2

Printed in Spain by Unigraf

Illustrations by IFA Design, Plymouth, UK

These resources have been developed to support teachers and students
undertaking a new OCR suite of GCSE Science specifications, Twenty First
Century Science.

Many people from schools, colleges, universities, industry, and the
professions have contributed to the production of these resources. The
feedback from over 75 Pilot Centres was invaluable. It led to significant
changes to the course specifications, and to the supporting resources for
teaching and learning.

The University of York Science Education Group (UYSEG) and Nuffield
Curriculum Centre worked in partnership with an OCR team led by Mary
Whitehouse, Elizabeth Herbert and Emily Clare to create the specifications,
which have their origins in the Beyond 2000 report (Millar & Osborne,
1998) and subsequent Key Stage 4 development work undertaken by UYSEG
and the Nuffield Curriculum Centre for QCA. Bryan Milner and Michael
Reiss also contributed to this work, which is reported in: 21st Century
Science GCSE Pilot Development: Final Report (UYSEG, March 2002).

Sponsors
The development of Twenty First Century Science was made possible by
generous support from:
- The Nuffield Foundation
- The Salters' Institute
- The Wellcome Trust

Homeostasis - Higher   B4

1 In and out of cells

Molecules move in and out of cells all the time.
This happens so that conditions inside the cell are kept steady.

a Complete the diagram of a cell to show chemicals moving in and out. Use words from the list.

- oxygen
- carbon dioxide
- urea
- digested food (e.g. glucose)
- water

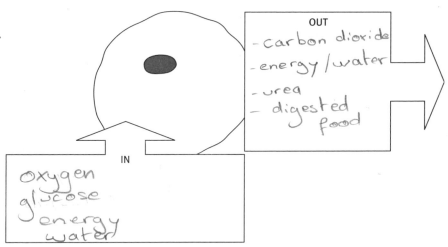

OUT
- carbon dioxide
- energy/water
- urea
- digested food

IN
oxygen
glucose
energy
water

b Use coloured pencils to underline the words you wrote in the boxes above.

- toxic waste products = **red**
- raw materials for respiration = **blue**
- raw materials for other chemical reactions in cells = **green**

Some words may be underlined in more than one colour.

---

2 Control systems

This incubator has an automatic control system.
It keeps conditions inside the incubator at steady levels.

a Label the diagram to show these parts of the control system:

| stimulus | receptor | processing centre |
| effector | response | |

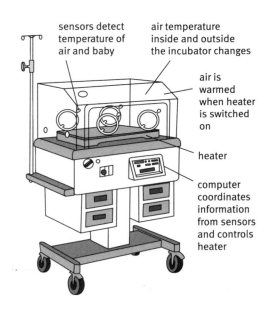

sensors detect temperature of air and baby

air temperature inside and outside the incubator changes

air is warmed when heater is switched on

heater

computer coordinates information from sensors and controls heater

b Your body also has control systems for homeostasis.

Complete the definition of homeostasis:

Homeostasis means _the maintenance of a constant internal environment_

c Body temperature is controlled by homeostasis.

Give one example of another factor that must be kept steady in the body.

water levels are required to be steady.

3

3 Negative feedback

a Write a definition for negative feedback.

→ negative feedbacks is used to decrease the temperature to a steady state.

b Explain why negative feedback is important in control systems.
Use the example of an incubator to help your explanation.

Because, negative feedback between the effector and the receptor of a control system reserves any changes to the systems steady state.

4 Antagonistic effectors

An incubator has only one effector. It has a heater to warm up the air.

Your body temperature control system has effectors to both:
- warm your body up *and*
- cool your body down

a Explain why body temperature control is an example of antagonistic effectors.

b Explain why having antagonistic effectors is an advantage in a body control system.

5 Balancing body temperature

Animals gain and lose heat from their environment.

Complete this sentence.

To keep a steady body temperature, input and output must be balanced so that:

heat ___gain___ = heat ___loss___

Homeostasis - Higher B4

6 Body temperature control systems

This text describes how your body controls its temperature.

Read the text then answer the questions.

> Receptors in your skin detect changes in the temperature of the air around you. Receptors in your brain detect changes in the temperature of your blood. This information is passed to the temperature control centre in the brain – called the hypothalamus.
>
> The hypothalamus coordinates all the information from the receptors. It automatically triggers effectors in the body to respond to changes in body temperature. These effectors are sweat glands and muscles.
>
> If body temperature rises too high:
> - sweat glands increase production of sweat
> - muscles in blood vessels supplying the skin relax
>
> These responses work to lower body temperature.
>
> If body temperature becomes too low:
> - sweat production is stopped
> - muscles in blood vessels supplying the skin contract
> - skeletal muscles contract rapidly, causing shivering
>
> These responses work to raise body temperature.

a Shade each of the boxes next to these terms a different colour.

stimulus ☐   receptor ☐   processing centre ☐   effector ☐   responses ☐

b Use these colours to highlight or underline parts of the temperature control system in the text above.

c Explain *how* these responses help to raise or lower body temperature:

- producing more sweat *to keep the body cool down*

- shivering *to help you get warm*

d Blood vessels carrying blood to the skin have muscles in the walls.
These muscles can be contracted (vasoconstriction) or relaxed (vasodilation).

The diagrams show vasoconstriction and vasodilation.
Explain *how* these responses help to control body temperature.

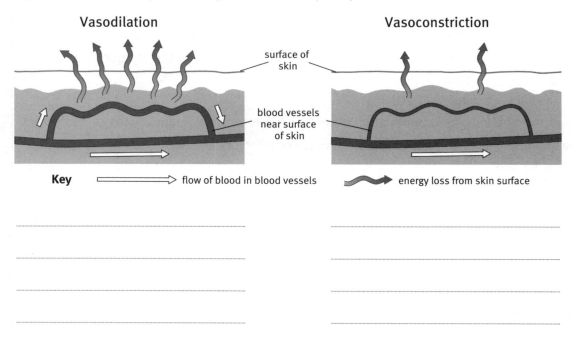

7 Enzymes

Enzymes are a very important group of chemicals found in living things.

a What type of chemical molecule are enzymes made of? ....

Hydrogen peroxide is a waste product of chemical reactions in many living cells. It is poisonous, so it is broken down by cells into water and oxygen. The enzyme catalase is very important in breaking down hydrogen peroxide.

A student measured how fast a sample of hydrogen peroxide was broken down with and without catalase.

The diagram shows their equipment. The graph shows their results.

b Where is the catalase in this experiment?

....

c What effect does catalase have on the breakdown of hydrogen peroxide?

....

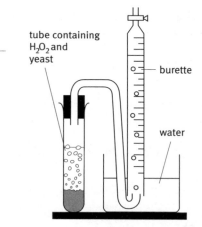

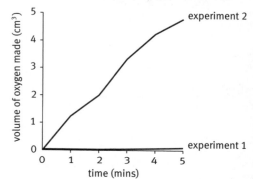

Homeostasis - Higher B4

d Complete the boxes to explain how the enzyme catalase works.

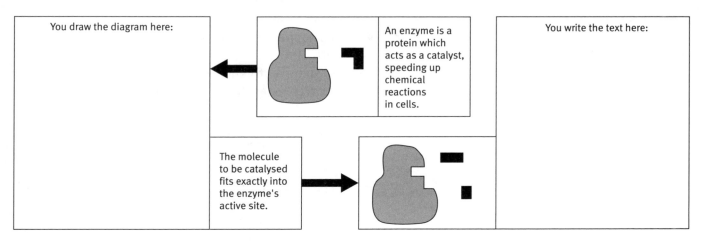

e There are tens of thousands of enzymes in the human body. Each one speeds up a different chemical reaction.

Explain why an enzyme can speed up only one particular reaction. Use these key words in your answer:

| active site | shape | substrate | lock and key |

---

8 Enzymes and temperature

Enzyme reactions are affected by temperature. Complete the notes to explain why this happens. Use these key words in your notes:

collisions   optimum   denatured

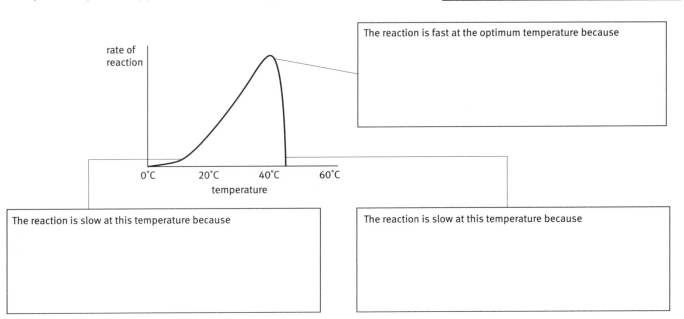

9 Diffusion

Molecules are always moving. In gases and liquids they move randomly.
The molecules bump into each other and spread out. This happens by diffusion.

a Tom was watching television when his mother sprayed some air freshener at the other side of the room. After a few minutes Tom could smell the scent of the air freshener. The diagrams explain what happened.

| The air freshener molecules were at a high concentration near the can after they had been sprayed. | → | The air freshener molecules moved about randomly. They collided with each other and other molecules in the air. | → | The air freshener molecules moved from their high concentration near the can to regions of lower concentration. | → | Some air freshener molecules reached Tom's nose, so he could smell the scent. |

b Tom then made himself a drink. He added some blackcurrant cordial to water in a glass.
After a few minutes, all of the water was the same purple colour.

Explain what happened in the flow chart below.

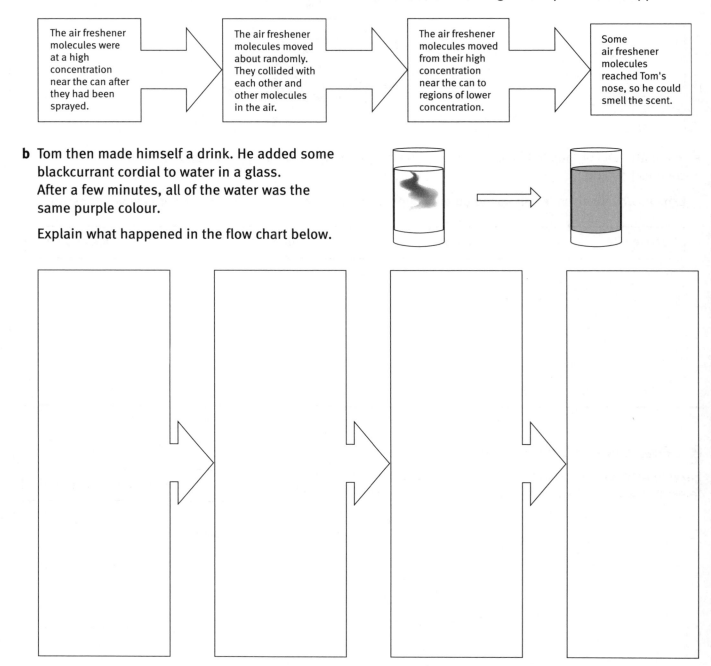

c Write down three examples of chemicals which move in or out of cells by diffusion.

10 Osmosis

Osmosis is a special case of diffusion.

a Write down a definition of osmosis.

....

....

b The diagram shows a semi-permeable bag filled with sugar solution. It is sitting in a beaker of a different concentration of sugar solution.

Draw arrows to show the movement of water molecules between the two solutions.

(Remember, water molecules will move in *both* directions. Use different sized arrows to show the *overall* movement of water molecules.)

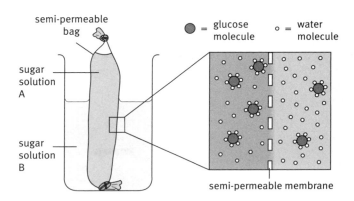

c Animal cells can be damaged if their water balance is upset.

A red blood cell was put in a very dilute sugar solution.

⇨ The solution was more dilute than the contents of the cell.

⇨ There was a higher concentration of dissolved molecules in the cell than in the solution.

⇨ Water entered the cell by osmosis, so the cell swelled up and burst.

A second red blood cell was put into a concentrated sugar solution.

Write three points to explain what has happened to this second cell.

⇨ ....

⇨ ....

⇨ ....

B4 Homeostasis - Higher

11 Active transport

a Complete the sentences to describe active transport. Use words from the list.

| active transport | high | low | energy | passive |

Diffusion does not need any _____ from the cell. It is a _____ process. Diffusion only moves molecules from a _____ to a _____ concentration. Some molecules move into cells by a process called _____ _____.

b Colour the numbered parts in the diagram to match the words in these sentences.

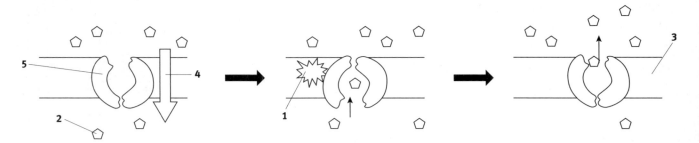

Active transport uses energy₁ to move molecules₂ across cell membranes₃ from low to high concentrations. This is against their concentration gradient₄. The energy is used to change the shape of a carrier protein₅ in the membrane.

---

12 Water balance

Water moves in and out of your body every day. These inputs and outputs must be balanced for your cells to work properly.

Complete the table to show how your body gains and loses water.

List ways your body gains water	List ways your body loses water
1. Eating	1.
2. drinking	2.
3.	3.

Homeostasis - Higher B4

13 Kidneys balance the body's water levels

The kidneys make urine. The amount of water in urine changes.

Complete the table to show how different conditions affect the concentration of your urine.

Conditions	Concentration of blood	Level of water in urine	Concentration of urine
cold day, staying inside	low	high	dilute
hot day, playing sport outside			
eating lots of salty food			
drinking lots of liquids			

14 Kidneys get rid of the body's waste

a Use these words to label the diagram on the left.

| urine | tubules | kidney |

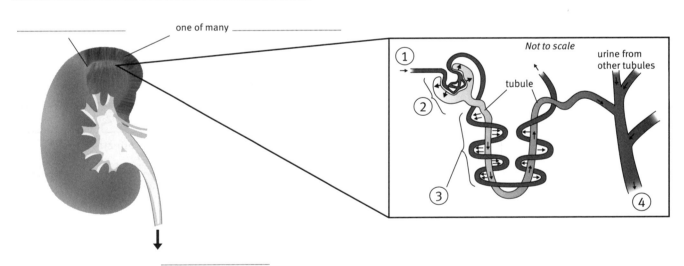

one of many

Not to scale

urine from other tubules

tubule

b Put the correct number in each box to describe what is happening in the diagram on the right.

☐ Small molecules are **filtered out** of the blood into the tubule, including urea, glucose and water. Salt ions are also filtered out.

☐ **Reabsorption** of useful ions of salt, glucose and water molecules.

☐ Blood flows to the tubule.

☐ Waste molecules form urine which is stored in the bladder.

15 Water balance is controlled by ADH

The amount of water in urine is controlled by a hormone called ADH.

a Complete the control system diagram to explain how this works.

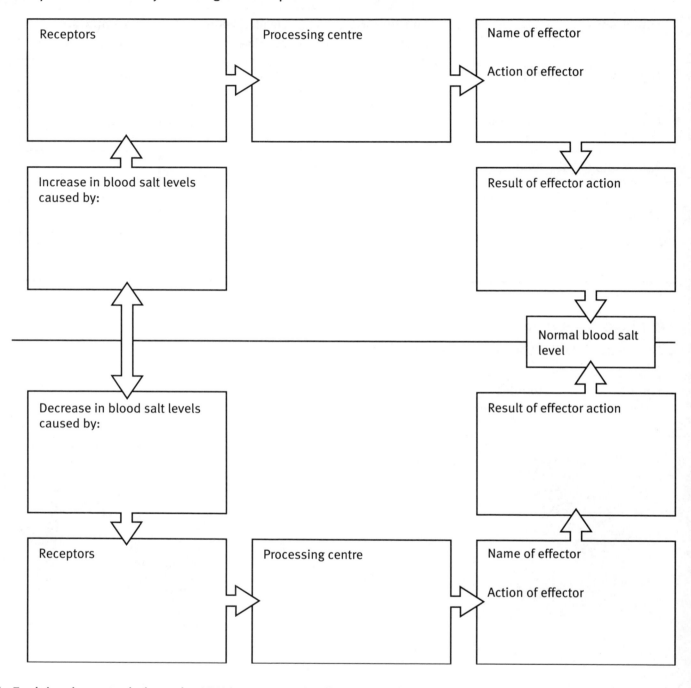

b Explain why water balance by ADH is an example of negative feedback.

....

....

....

Homeostasis - Higher   B4

16 Extreme environments

a Complete the list of four factors which must be kept steady in the body:

temperature _____

_____ _____

b Complete the graph to show the normal range of core body temperature.

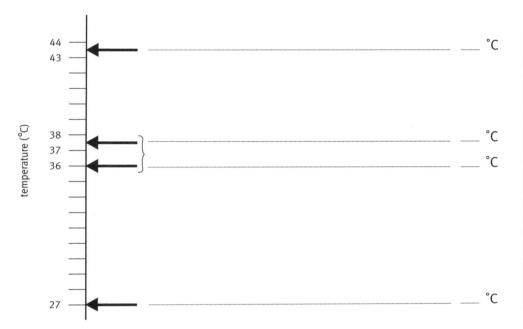

c Write down the parts of the body that make up the core.

_____

d Write down two parts of the body that are extremities.

_____ and _____

e Very harsh conditions can stop homeostasis from working properly. Complete the sentences to explain how very hot temperatures can affect the body. Use words from the list.

| core | decrease | up | rise | low | increases | dehydrated |

Condition: Very hot weather.

Homeostasis demands: Body temperature goes _____. Sweating _____.

This may cause body water levels to fall too _____ if the lost water is not replaced.

The person may become _____. If this happens sweating may _____.

This will cause body temperature to _____ even higher. If the _____ body temperature becomes too high, the body's temperature control systems stop working.

13

B4 Homeostasis - Higher

f Explain how very cold weather can affect the body's homeostasis systems.

Condition: Very cold weather.

Homeostasis demands: _____

_____

_____

_____

g Explain how two different sports can put demands on the body's homeostasis systems.

Condition: _____

Homeostasis demands: _____

_____

_____

_____

Condition: _____

Homeostasis demands: _____

_____

_____

_____

Homeostasis - Higher **B4**

17 Heat stroke

a Complete the definition of heatstroke.

Heatstroke is ＿＿

b Complete the information leaflet about heatstroke.

Avoiding heatstroke

People may suffer heatstroke as a result of

→ ＿＿

→ ＿＿

Look out for the following symptoms:

→ ＿＿

→ ＿＿

→ ＿＿

You should take the following actions to cool the patient:

→ ＿＿

→ ＿＿

→ ＿＿

→ ＿＿

B4 Homeostasis - Higher

18 Hypothermia

a Complete the definition of hypothermia.

Hypothermia is when _____ .

Body heat cannot be _____ .

b Complete the information leaflet about hypothermia.

Avoiding hypothermia

People on mountains may become chilled as a result of

⇒ _____

⇒ _____

Look out for the following symptoms:

⇒ _____

⇒ _____

⇒ _____

⇒ _____

You should take the following actions to warm the patient:

⇒ _____

⇒ _____

⇒ _____

⇒ _____

Do not give a hot water bottle because

⇒ _____

Do not give food or alcohol because

⇒ _____

Homeostasis - Higher B4

19 Drugs and homeostasis

Certain drugs can upset the body's water level control system.
They do this by affecting the production of ADH.

a Complete the sentences to explain the role of ADH. Use words from the list.

| increases | more | decreases | reabsorbed | pituitary | kidney | hormone |

Water is filtered out of the blood in the _____. Some of this water is _____ to keep the body's water level balanced.

ADH is a _____. It is made by the _____ gland in the brain. ADH causes the kidneys to reabsorb _____ water.

When the body's water levels are too low, ADH production _____.

If the body's water levels are too high, ADH production _____.

b Complete the table to show the effects that certain drugs have on the body's water balance.

Drug	Effect on ADH production	Urine production	Effect on body water balance
caffeine	decreased	greater volume of dilute urine made	dehydration
alcohol			
Ecstacy			

c Ecstacy also increases sweat production. Explain how this may cause serious disruption to the body's temperature control system.

_____

_____

Chemical patterns - Higher   C4

1 The elements in order

Complete the sentences and diagram.

Scientists look for patterns in data. When they arrange the known elements in order of relative _____ _____, they find that there is a repeating pattern. These patterns are shown clearly when the elements are arranged in a _____ _____. Each row in the table is a called period, with metals on the _____ and _____ on the right.

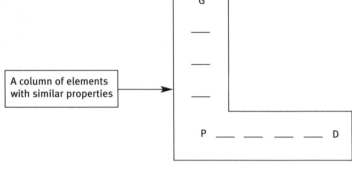

A column of elements with similar properties

Vertical columns of elements in the table are _____ of elements with similar properties.

Modern versions of the periodic table put the elements in order of _____ number, also known as the atomic number.

2 The periodic table

a Colour the key, then use the colours to show these parts on the periodic table below.

☐ group 1 (alkali metals) ☐ group 7 (halogens) ☐ non-metals ☐ transition metals

b Over three-quarters of the elements are metals. Lightly colour in the metals.

[Periodic table]

18

3 Periodic patterns

This is a plot of melting point against proton number for some elements.

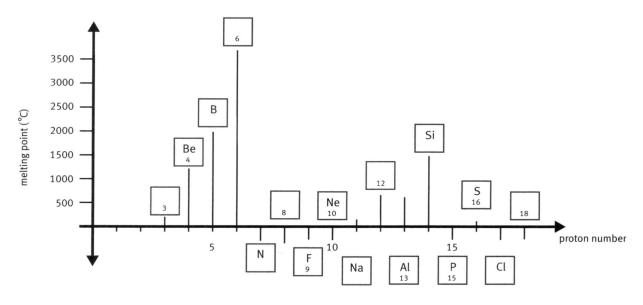

a Use the periodic table on page 18 to fill in the five missing proton numbers and five missing symbols.

b Which two elements are at the peaks? _____

   To which group do these two elements belong? _____

c Describe the pattern of melting points across the two periods.

_____

This is a plot of melting point against proton number for a group of elements.

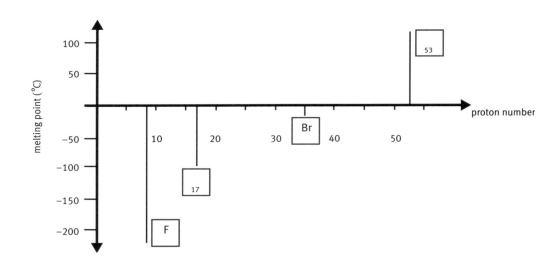

d Use the periodic table on page 18 to fill in the two missing proton numbers and two missing symbols.

e To which group do these two elements belong? _____

f Describe the pattern of melting points down the group.

_____

C4 Chemical patterns - Higher

4 The alkali metals

a Complete the information about group 1 elements. The alkali metals:

- are .... – you can cut them with a knife
- are .... – but only when freshly cut
- quickly .... in moist air – they react with water and oxygen
- have low .... – some float on water
- react with water to form .... and an .... solution of the metal ....

e.g. sodium + water → sodium hydroxide + hydrogen

lithium + ... → .... + ...

We call group 1 the **alkali** metals because they produce alkaline solutions when they react with water. They also react vigorously with chlorine. The products are **colourless** crystalline salts called metal chlorides.

e.g. sodium + chlorine → sodium chloride

potassium + ... → ....

b Explain the precautions you must take when using group 1 metals and alkalis.

....

....

c From the information in the table:

Alkali metal	Proton number	Melting point (°C)	Boiling point (°C)	Density (g/cm³)
lithium	3	181	1331	0.54
sodium	11	98	890	0.97
potassium	19	63	766	0.86

- How does melting point vary with proton number in group 1?

....

- How does boiling point vary with proton number in group 1?

....

- What further information do you need to decide whether or not there is a pattern to the densities of group 1 metals?

....

Chemical patterns - Higher C4

d In the table below, describe the reactions of alkali metals with water (right-hand column). Choose words from this list.

violent reaction	makes sparks	gas catches fire	
moves around on water	fizzes gently	makes sparks	melts
metal thrown off surface	floats	gives off hydrogen	

Write these words into the left-hand column to show the trend in reactivity down the group:

most reactive	least reactive

Reactivity	Name of metal	Description of reaction with water
	lithium	
	sodium	
	potassium	

5 Names and formulae

Complete the table, with the help of the periodic table on page 18.

Chemical	Symbols of the element(s) in it	Formula of the chemical
hydrogen	H	H_2
water	........., ...	
lithium fluoride	Li, ...	
sodium chloride	Na, ...	
potassium bromide	K, ...	
lithium hydroxide	Li, ..., ...	

6 Balanced equations

a Use these words and symbols to complete the sentences.

| g | aq | s | arrow | l | balanced | atoms |

In a symbol equation, the number of _____ of each element on each side of the _____ in an equation must be equal. We call this a _____ equation.

We can also show the reactants and products as solid (____), liquid (____), gas (____), or aqueous solution (____).

b Complete the equations to describe the reaction between potassium and water.

Step 1: Describe the reaction in words:

potassium + water → _____ + _____

Step 2: Write down the formulae for the reactants and products:

_____ + _____ → _____ + _____

Step 3: Balance the equation:

_____ + _____ → _____ + _____

Step 4: Add state symbols:

_____ + _____ → _____ + _____

c Complete the equations to describe the reaction between lithium and chlorine.

Step 1: Describe the reaction in words:

lithium + chlorine → _____

Step 2: Write down the formulae for the reactants and products:

_____ + _____ → _____

Step 3: Balance the equation:

_____ + _____ → _____

Step 4: Add state symbols:

_____ + _____ → _____

7 The halogens

a Complete these sentences by drawing a ring around the correct **bold** words.

- The halogens are **metals / non-metals** and their vapours are **coloured / colourless**.
- Halogens can **colour / bleach** vegetable dyes and kill bacteria.
- The halogens are **toxic / non-toxic** to humans.
- Halogen molecules are each made of **one / two** atoms; they are **monatomic / diatomic**.
- Halogens react with **metal / non-metal** elements to form crystalline compounds that are salts.
- The halides of alkali metals are **coloured / colourless** salts such as **potassium / iron bromide**.
- Compounds of group 1 elements have the formula **MX / MX$_2$** (where M = metal and X = halide).

b Complete the table. Choose the missing temperatures in the second and third rows from these values:

| −34 °C | 58 °C | 114 °C |

Name	chlorine	bromine	iodine
State at room temperature			
Melting point (°C)	−101	−7	
Boiling point (°C)			184

c Describe the reactions of these halogens with hot iron. Draw an arrow to show decreasing reactivity down the group.

Reactivity	Name of metal	Description of the reaction with hot iron
	chlorine	
	bromine	
	iodine	

8 Flame colours and spectra

a Colour these flames to show the results of a flame test with each of the named salts.

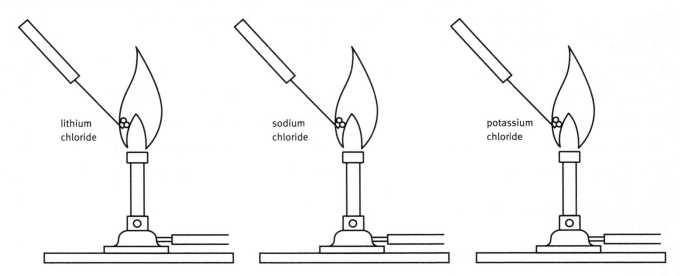

b Colour the lines in this spectrum of helium gas.

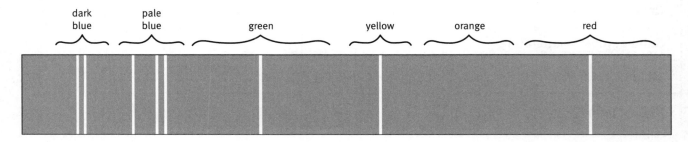

c How does the spectrum of helium differ from the spectrum of white light from the Sun?

d Why is it possible to use spectra to identify elements during chemical analysis?

e How was it possible to discover helium on the Sun before it was discovered on Earth?

9 Atomic structure

a All atoms are made of the same basic parts – protons, neutrons, and electrons.

Label the diagram of an atom.

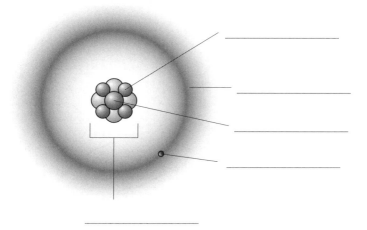

b Complete the table.

Part of atom	Relative mass	Charge	Position
proton		+1	in a very small central nucleus
	1		
electron			in shells around the atom's nucleus

c Complete the sentences using these words.

| electrons | proton | protons | charge | atomic |

↳ All atoms of the same element have the same number of ………………………… .

↳ The number of protons is called the ………………………… number or ………………………… number.

↳ The number of protons is equal to the number of ………………………… . This means that an atom has no overall ………………………… .

10 Electrons in atoms

a Complete these sentences by filling in the blanks with words or numbers.

The electrons in an atom are arranged in a series of _____ around the nucleus. These shells are also called _____ levels. In an atom the _____ shell fills first, then the next shell, and so on.

There is room for

- up to _____ electrons in shell one
- up to _____ electrons in shell two
- up to _____ electrons in shell three

Shells fill from _____ to _____ across the _____ of the periodic table.

- Shell one fills up first from _____ to helium.
- The second shell fills next from lithium to _____.
- Eight _____ go into the third shell from sodium to argon.
- Then the fourth shell starts to fill from potassium.

The _____ number of sodium is 11. So there are _____ electrons in a sodium atom.

The diagram above shows the arrangement of _____ in a sodium atom. This electron arrangement can also be written: _____.

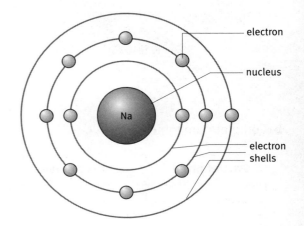

b Show the arrangements of electrons in these atoms. Sodium has been done for you. You will find the proton numbers in the table on page 18.

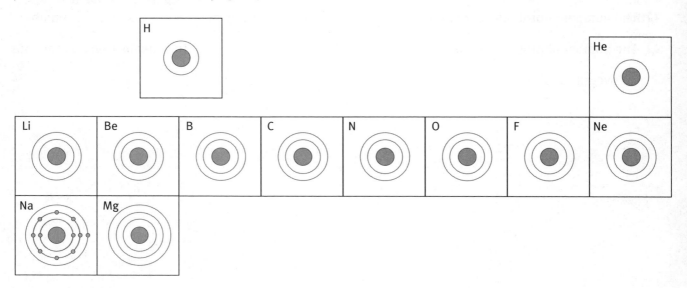

c Complete the lists of the arrangement of electrons (electronic configurations) for:

Alkali metals (group 1 in the periodic table)		**Halogens** (group 7 in the periodic table)	
lithium	2.1	fluorine	
sodium		chlorine	
potassium		bromine	2.8.18.7

d Since it is the electrons in the outer 'shell' that affect chemical reactions, the number of outer-shell electrons determines the chemical properties of an element.

Complete these sentences.

➔ Group 1 metals have similar properties because they have .... electron in their outer shell.

➔ Halogens have .... properties because they have .... electrons in their outer shell.

e Give examples or more details to illustrate or justify the general statements made in these two paragraphs.

When atoms react it is the electrons in the outer shell which get involved as chemical bonds break and new chemicals form. Elements have similar properties if they have the same number and arrangements of electrons in the outer shells of their atoms.

The alkali metals are not all the same because their atoms differ in the number of inner full shells. A sodium atom has two inner filled shells, so it is larger than a lithium atom and its outer electron is further away from the nucleus. As a result, the two metals have similar but not identical physical and chemical properties.

11 Salts and their properties

Complete this table to give examples to illustrate the properties of salts.

Description of salts	Example
Salts are compounds of metals with one or more non-metals.	
Salts are crystalline when solid.	
Salts have much higher melting and boiling points than compounds made up of small molecules.	
Some salts are soluble in water.	
Some salts are insoluble in water.	
Salts do not conduct electricity when solid but they do when molten (liquid). There are changes at the electrodes when an electric current flows through a molten salt.	
Solutions of salts in water conduct electricity. There are changes at the electrodes when an electric current flows through a solution of a salt.	

12 Ionic theory

a The diagram shows a model of the structure of sodium chloride.

Colour the diagram to show:
- sodium ions **red**
- chloride ions **green**

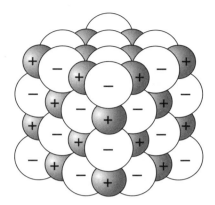

How does this model of the structure of sodium chloride explain the shape of sodium chloride crystals?

_____

_____

_____

b Complete this table.

Property of salts	Ionic theory explanation
All the crystals of each solid ionic compound are the same shape. Whatever the size of the crystal, the angles between the faces of the crystal are always the same.	
	The giant ionic structure is held together by the strong attraction between the positive and negative ions. It takes a lot of energy to break down the regular arrangement of ions.
	In a molten ionic compound the positive and negative ions can move around independently.
The solution of an ionic compound in water is a good conductor of electricity.	

13 Atoms into ions

a Complete the sentences next to these diagrams.

Atoms of metals on the left-hand side of the periodic table turn into ions by losing electrons.

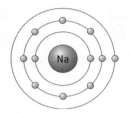 When it turns into an ion the sodium atom _____ 1 electron (negative charge)

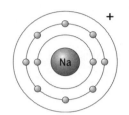 so the sodium ion has a _____ charge

Atoms of non-metals to the right of the periodic table turn into ions by gaining electrons.

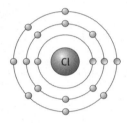 When it turns into an ion the chlorine atom _____ 1 electron (negative charge)

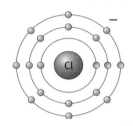 so the chloride ion has a _____ charge

b Complete these diagrams to show the number and arrangement of electrons in each atom and the ions they form. (Your answers to question 10b will help you.)

Atom   **Ion**

Symbol: Li   Symbol: Li$^+$

Symbol: Mg      Symbol: _____

Symbol: F   Symbol: _____

Symbol: O   Symbol: _____

14 Formulae of ionic compounds

This table shows formulae of simple ions.

		H^+								
Li^+							N^{3-}	O^{2-}	F^-	
Na^+	Mg^{2+}				Al^{3+}			S^{2-}	Cl^-	
K^+	Ca^{2+}					no simple ions			Br^-	no ions formed
Rb^+	Sr^{2+}	transition metals form more than one ion, e.g. Fe^{2+}, Fe^{3+}							I^-	
Cs^{2+}	Ba^{2+}									
1+	2+				3+		3−	2−	1−	

metals — positive ions   non-metals — negative ions

a Complete these general statements about ions.

↪ The metals in group 1 and 2 form _____

↪ The alkali metals form ions with _____

↪ Non-metals in groups 6 and 7 form _____

↪ The halogens form ions with _____

b Complete this table to show that ionic compounds are electrically neutral overall because the positive and negative charges balance.

Ionic compound	Ions present		Formula
	Positive ions	**Negative ions**	
	Li^+		LiBr
magnesium iodide	Mg^{2+}	I^- I^-	
			$AlBr_3$
sodium oxide			

15 Chemical hazards

Complete the descriptions of chemical hazards by filling in the blanks. Then write the names of these chemicals alongside the matching hazards. Some chemicals have more than one hazard.

| chlorine | bromine | iodine | sodium | sodium hydroxide |
| potassium nitrate | sodium carbonate | dilute | hydrochloric acid |

Symbol	Hazard	Examples
✖	**Harmful** A chemical which may involve limited health risks if breathed in, _____ or taken in through the skin. (Less hazardous than a toxic chemical.)	
☠	**Toxic** A chemical which may involve serious health risks or even _____ if breathed in, swallowed, or taken in through the _____	
✖	**Irritant** A chemical which may cause sores or inflammation in contact with skin or eyes. (Less hazardous than a _____ chemical.)	
(corrosive symbol)	**Corrosive** A chemical which can destroy living tissues such as _____ or _____.	
(flame symbol)	**Highly flammable** A chemical which may easily catch _____ or which gives of a flammable gas in contact with water.	
(flame over circle)	**Oxidizing** A chemical which reacts strongly with other chemicals and makes the mixture so hot that it may cause a _____.	

Explaining motion - Higher

P4

1 Interactions and forces

Forces happen because of an **interaction** between two objects. They happen in **pairs**.

a In each of the bullet points below, draw a ring round the correct **bold** word.

The two forces in an interaction pair

- are **always** / **sometimes** / **never** the same size
- act in **the same** / **random** / **opposite** directions
- act on **the same** / **a different** object

b In each of these drawings, add an arrow in red to show the named force.

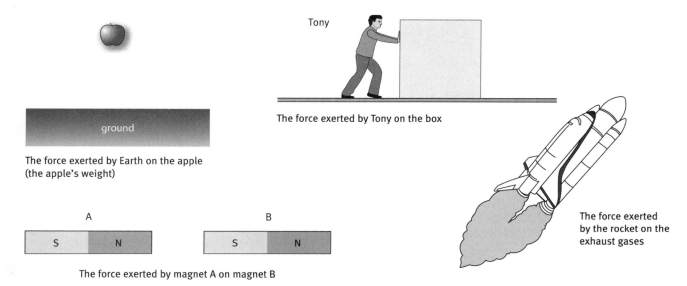

c Complete these sentences by filling in the missing words. Each sentence will help you with the next one.

i Magnet A pulls magnet B to the left; magnet B pulls magnet B to the _____ with the same size force.

ii Tony pushes the box to the right; the box exerts a force on Tony. It pushes him to the _____ with the _____ size force.

iii The Earth exerts a downwards force on an apple; the apple exerts an _____ force on the _____. It pulls back with the _____.

iv The Shuttle exerts a force backwards on the exhaust gases; the exhaust gases

_____

d Now add a second arrow to each of the pictures above. This should show the second force in the interaction pair. Use a different colour and label the force like this:

'force exerted by _____ on _____'.

33

P4 Explaining motion - Higher

2 Getting moving

a Look at the picture on the right. It shows Sophie pulling on a rope which is attached to a wall.

Draw a circle round two places where friction is helping her grip.

Sophie

b A bicycle tyre grips the road. When the wheel turns, the bottom of the tyre pushes backwards on the ground. In the picture, the tyre exerts a force to the left on the ground.

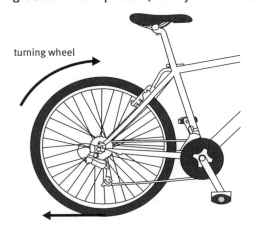
turning wheel

 i What is the force that allows the tyre to push backwards on the ground? _____

 ii The ground exerts a force on the tyre. What is the direction of this force? _____

 iii Draw an arrow on the wheel to show the direction and size of this force.

 iv Label the two arrows on the picture.

c Use words from the list to complete the sentences below.

Words may be used once, more than once, or not at all.

| upwards | downwards | forwards | backwards | grip | rope | exerts |

➔ A racing car accelerates at the start of a race: friction allows the tyres to _____ the track.

 The tyres exert a force backwards on the track; the track _____ a force _____ on the tyres. This makes the car speed up.

➔ A climber uses a rope to go up a mountain: friction allows her hands to grip the _____ .

 Her hands exert a _____ force on the rope; the rope _____ an _____ force on her hands.

➔ Mick is trying to walk to the shops. The ground is icy and slippery.

 His feet cannot get a _____ to push _____ on the surface.

 So the surface does not push him _____ .

34

Explaining motion - Higher **P4**

3 How surfaces hold things up

a The picture shows a bag on a springy cushion.

  i How can you tell that the bag is exerting a force on the cushion?

  ....

  ii Draw a red arrow to show the force that the bag exerts on the cushion. Label this arrow 'force exerted by bag on cushion'.

  iii The cushion reacts by pushing up on the bag.
  How does the cushion exert a force on the bag?

  ....

  iv What is the name of this force? ....

  v Draw a second arrow to show the force the cushion exerts on the bag.
  Use a different colour and label this force with its name.

  vi What would happen to the cushion if the bag were removed?

  ....

b The picture shows an apple resting on a table. Its weight is pulling it down. But the apple is not falling.

  i What is the force (and object) that stops the apple from falling?

  .... from the ....

  ii Does the table get squashed by the apple? ....

  iii Explain why we don't see the table being squashed.

  ....

  iv Explain how the table exerts a force on the apple.

  ....

c Complete this sentence to describe the forces between an object and a surface it is resting on.

When an object rests on a surface, it exerts a force .... on the surface.

The surface exerts a force .... on the object.

This is called a .... force.

4 Weight and reaction

So far, you have looked at pairs of forces acting *between* two different objects. Now you will work with more than one force acting on a *single* object.

a The picture shows a tennis ball on the ground. Its mass is 0.2 kg.

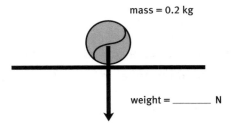

mass = 0.2 kg

weight = ... N

i What is its weight? (Take the gravitational field strength as 10 N/kg.)

weight = ... N

ii Write this value on the picture.

iii Draw an arrow to show the reaction force exerted by the ground on the ball.

iv Label this force.

v There are two forces acting on the ball.

What is the **resultant** force acting on the ball? ....

b The picture shows a mass hanging on a thread. Its weight is 4 N.

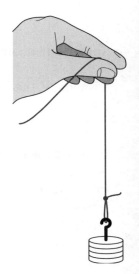

i Draw arrows for the two forces that are acting on the mass.

Use these phrases to label the forces:

- force exerted by the Earth on the mass
- force exerted by the thread on the mass

ii What is the resultant force on the mass? ....

iii Imagine the thread breaks. Put a cross ✗ next to the force that will disappear.

iv Put a tick ✔ next to the force that will still be there after the thread breaks.

v Immediately after the thread breaks, what is the resultant force on the mass?

....

c Clare and Sophie have a tug-of-war. They pull the rope with the same amount of force. They are not moving.

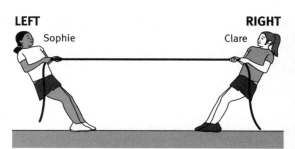

LEFT — Sophie   RIGHT — Clare

i Draw and label arrows on the diagram to show:

- force exerted by Clare on the rope
- force exerted by Sophie on the rope

ii What is the resultant horizontal force acting on the rope?

....

36

Explaining motion - Higher   **P4**

5 Average and instantaneous speed

a The average speed of an object is:

$$\text{speed (m/s)} = \frac{\text{distance travelled (m)}}{\text{time taken (s)}}$$

In the 1995 World Athletic Championships, Linford Christie was the defending champion, in the 100 m sprint. He lost to Donovan Bailey, who won with a time of 9.97 seconds.

Calculate Bailey's average speed for the race.

Bailey's average speed = ... m/s

b The table below shows the time six of the athletes took to complete the race.

Finishing position	Athlete	Finishing time in s	Average speed in m/s
1	Bailey	9.97	
2	Surin	10.03	9.97
3	Boldon	10.03	
4	Fredericks	10.07	
5	Marsh	10.10	
6	Christie	10.12	

i Complete the table by calculating the average race speed for the athletes. The second one has been done for you.

ii What happens to the speed as the time gets longer? ....

c i How long would it take for Surin to run 800 m at this average speed?

time for Surin to run 800 m = ... s

ii Complete these sentences by drawing a (ring) around the correct **bold** words.

In reality, if Surin were to keep running for 800 m, his average speed would be **more / less** than 9.97 m/s. So the time he would take is **longer / shorter** than the answer above.

37

P4 Explaining motion - Higher

Where speed is changing, the average speed and instantaneous speed can be different.

d The following diagram shows how the positions of six of the runners changed at each 25 metre mark.

Time:	0.00 s	3.00 s	5.50 s	8.10 s	9.97 s
Distance:	0 m	25 m	50 m	75 m	100 m
1: Christie		●	●	●	
2: Fredericks		●	●	●	●
3: Surin		●	●	●	●
4: Boldon		●	●	●	●
5: Bailey		●	●	●	●
6: Marsh		●	●	●	

i Which athlete got the best start to the race (at 3.00s)? _____

ii Use the information in the diagram to calculate Bailey's average speed over the last 25 m of the race.

Bailey's average speed for last 25 m = _____ m/s

iii Complete these sentences using words from the box. There is one word you don't need to use and one word you need to use twice.

instantaneous	average	lower	higher	speed

Bailey's speed at any one moment is called his _____ speed. Just after the start,

his instantaneous _____ is lower than it is when he gets going. As a result, his

_____ speed for the last 25 m is _____ than his _____

speed for the first whole race.

e During the race, Christie suffered a hamstring injury.

Look carefully at the pictures and decide where you think Christie's injury happened. Explain your answer.

Explaining motion - Higher P4

6 Using graphs to summarize motion

a Look at the distance–time graph on the right.

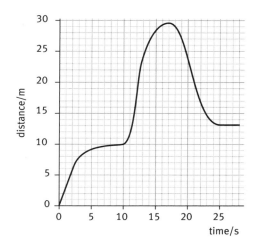

  i Label the graph with these letters to show where the object is
     A moving away from the start with a constant speed
     B moving backwards towards the start with a constant speed
     C stationary

  ii You can tell from the graph that the object moved faster between 0 and 5 seconds than it did between 10 and 15 seconds. Explain how you can see this from the graph.

  ....

  iii Work out the speed of the object between 10 seconds and 15 seconds.

  speed = ... m/s

Straight slopes on a distance–time graph tell you that an object is moving at a constant speed.
Curves tell you that an object's speed is **changing** – it is speeding up or slowing down.

b Label the graph with these letters to show where the object is
  D slowing down (going forwards)
  E speeding up (going forwards)
  F stationary
  G slowing down (going backwards)
  H speeding up (going backwards)

A graph of speed against time is another useful way of showing how an object moves.

c Look at the distance–time graphs (left) and the speed–time graphs (right) below.

  Draw lines to match each distance–time graph with its corresponding speed–time graph. The first one has been done for you.

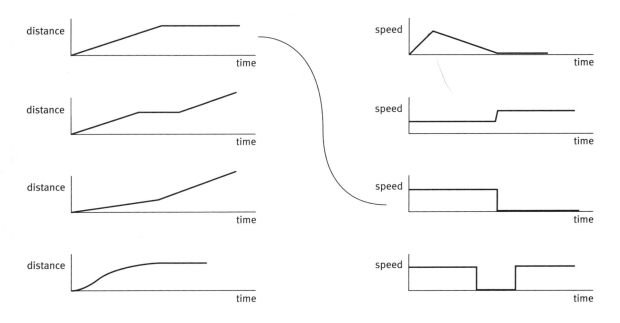

P4 Explaining motion - Higher

A tachograph is a speed–time graph to record a lorry's journey.

There are strict rules to control how long a lorry driver can drive without taking a break. A tachograph is used to record the lorry's journey.

d i What was the highest instantaneous speed of the lorry?

_____

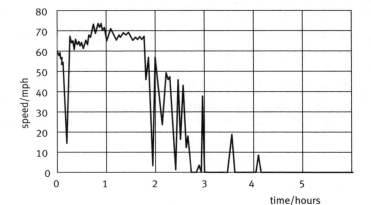

ii During the journey the lorry
- **A** drove in a town
- **B** drove on the motorway
- **C** parked up for the night
- **D** made some deliveries

Use the letters to label sections of the graph that show these actions.

iii The average speed of the lorry in the first two hours was 62 mph.

Draw a coloured line to show this on the graph.

iv In the first two hours, mark a time where the instantaneous speed was much less than the average speed.

v Suggest what might have happened at that moment. _____

A tachograph shows the **speed** of a lorry. It is a speed–time graph. You cannot tell which direction it was going.

A **velocity–time** graph gives a fuller description. It shows the **speed** and **direction**. Velocity means the speed in a certain direction.

e The velocity–time graphs below show the motion of a car travelling along a straight road running north–south.

Draw lines to match each graph with the best description of the motion. One has been done for you.

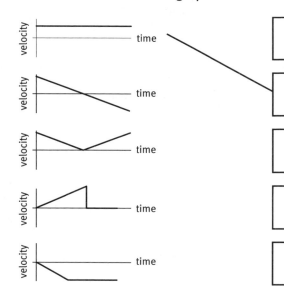

	The car travels north, steadily slowing down until it stops momentarily and then gradually speeding up again.
	The car is travelling north at a constant speed.
	The car gradually speeds up as it travels north and then suddenly comes to a stop.
	The car gradually speeds up as it travels south, eventually travelling at a constant speed.
	The car, which is travelling north, slows down and comes to a stop. It then travels south, getting faster and faster.

f Complete this sentence: When the car is moving backwards, its velocity is _____.

Explaining motion - Higher P4

7 Momentum

momentum	=	mass × velocity
(kg m/s)		(kg) (m/s)

a Look at the three balls on the right.
They are travelling at different speeds.

  i Calculate the momentum of the golf ball.

  ii Use the same method to calculate the momentum of the tennis ball and football.

  iii Put a tick ✔ next to the ball with the biggest momentum.

  iv Each of the balls strikes a large skittle. Put a cross ✘ next to the ball which is most likely to knock it over.

  v Explain your choice in part **iii**.

  ....

  ....

  ....

  ....

0.05 kg
→ 30 m/s

momentum = mass × velocity
= .... × ....
= .... kg m/s

0.05 kg
→ 45 m/s

momentum = mass × velocity
= .... × ....
= ....

0.5 kg
→ 4 m/s

momentum = ...

b Jo is ice skating. To get her moving, Sam gives her a push.

Work out Jo's momentum after she is pushed.

Momentum = ... × ....

= .... kg m/s

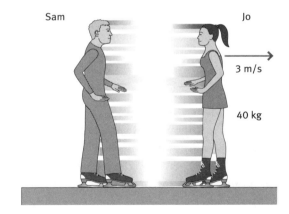

c Use words from the list to complete the sentences below.
Words may be used once, more than once or not at all.

| zero | momentum | force | left | right | same | opposite |

Jo is stationary before Sam pushes her. Her momentum is .... Sam exerts a

.... on Jo, pushing her to the .... The force acts for a short time;

as a result, her .... changes. She gains momentum going to the right – the

.... direction as the force that made this happen.

41

8 Forces and change of momentum

a Sam and Jo are the skaters from question **7b**.

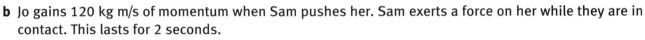

Before Sam pushes Jo, she is standing still. Her speed is 0 m/s. Sam exerts a force on Jo which increases her momentum. Sam could increase Jo's momentum more if he

➔ pushes her with more **force** or

➔ pushes her for a longer **time**

Write in the missing quantities and units to complete the equation:

Change of momentum = ... × ....

(..............................)   (...............) (...............)

b Jo gains 120 kg m/s of momentum when Sam pushes her. Sam exerts a force on her while they are in contact. This lasts for 2 seconds.
What is the size of the average force that Sam exerts on Jo?

Sam's force = ... N

c Lucie and Richard are on their skateboards. The ground is horizontal and they are not moving.

Lucie has a heavy ball. Lucie throws the ball and Richard catches it.

i This sentence describes what happens to Richard and Lucie. Put a ring around the correct **bold** words to complete the sentences.

Richard moves to the **left** / **right**; Lucie moves to the **left** / **right**.

ii Look at these phrases. They describe the exchange in terms of forces. But they are out of sequence. Put numbers in the boxes next to each phrase, to make sentences that show the correct sequence.

☐ To catch the ball, Richard exerts a force on it.

☐ the ball exerts a force on Lucie to the left.

☐ the force pushes the ball to the left.

☐ the force pushes the ball to the right;

☐ The ball exerts a force on Richard to the right.

☐ To throw the ball, Lucie exerts a force on it:

9 Changing momentum safely

a Look at the pictures of two crashed cars. Both cars were travelling at 20 m/s when they hit the wall. Look at these four phrases. They refer to the force on the driver as he is brought to rest.

  i Draw lines to link each of the phrases with one of the pictures.

  small time

  big time

  small force

  big force

  ii The driver's mass is 60 kg.

  What is his momentum before the collision?

  driver's momentum = _____ kg m/s

  iii He is brought to rest by the force from the seatbelt. The times are shown in the pictures.

  Use the equation on page 42 to calculate the force on the driver in

  ⇨ car **A**

  ⇨ car **B**

  iv It took ten times longer to stop the driver in car A. What effect did this have on the force acting on him?

b Imagine the driver in car B had not been wearing a seatbelt.

  i What object (or objects) would have provided the force to slow him down?

  ii It is safer to have *both* a seatbelt *and* a crumple zone. Explain the disadvantages of

  ⇨ a seatbelt on its own: _____

  ⇨ a crumple zone on its own: _____

10 Steady motion requires no (resultant) force

This is the **first law of motion**:

> If the resultant force acting on an object is zero, its momentum will stay the same.

a Complete these sentences. Draw a ring round the correct **bold** words.

This means that a **stationary** / **weightless** object will stay where it is.

It **will** / **will not** start moving if there is no resultant force.

A moving object will carry on at **the same** / **a lower** speed.

In both cases the momentum **does** / **does not** change because the resultant force is **zero** / **small**.

b The picture shows Georgia on her bicycle. Imagine Georgia increases her driving force to 150 N.

   i What will be the resultant force be on the bike?

      _____ N to the _____ .

   ii What will happen to her momentum?

      It will _____ to the _____ .

c Complete this sentence, which is part of the **second law of motion**:

> When a _____ force acts on an object, its _____ will change in the _____ direction as the force.

d The pictures show forces on some toy cars. Cars A and B are stationary. The others are moving to the right.

Fill in the blanks for the resultant force and its direction. Then draw lines to match each of pictures with a correct description of the car's motion. One has been done for you.

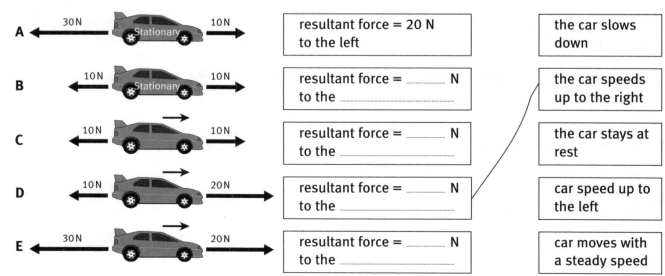

11 Work and change of energy

Whenever you move an object, you have to do some work.

The amount of work depends on
- the force you use
- the distance the object moves

a The equation for the amount of work done is:

work done (.........) = force (.........) × distance (.........)

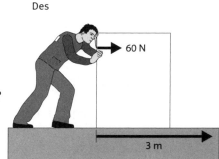

Des

i Complete the equation by putting the units in the brackets.

ii Des pushes a box 3 m across the floor. How much work does he do?

work done = ... N × .... m

= .... J

iii Energy is always conserved.

Complete these sentences to describe what happens to the work that Des does.

When Des pushes the box across the floor, the bottom of the box .... up. The work goes into .... the temperature of the box and the ground. When work is done on an object (or collection of objects), it increases the amount of .... stored in them.

b Gravitational potential energy (gpe)

Whenever you lift an object upwards, you have to do some work.
You are lifting against the force of gravity, which is pulling downwards on the object.

The object then has the potential to do the work back for you.

i Complete this sentence.

As a result of lifting the object up, it gains gravitational .... energy (gpe).

ii You lift a box that weighs 200 N on to a table that is 0.5 m high.

How much gravitational potential energy (gpe) has the box gained?

gpe gained = ... N × .... m

= .... J

iii Write down the equation you use to work out change in gravitational potential energy. Include the units.

Change in gpe = ... × ....

(.........) (.........) (.........)

P4 Explaining motion - Higher

12 Kinetic energy

a Complete this sentence.

You can do work on an object to make it speed up.

As a result, the object gains _____

_____ (KE).

Chris

mass of Chris and bicycle = 110 kg

speed = 8 m/s

distance travelled while speeding up = 50 m

forward force exerted by Chris = 100 N

b Chris does work at the start of a race to get her bike going.
From a standing start, she pedals along a straight road and reaches a speed of 8 m/s.

The mass of Chris with her bike is 110 kg.

i Write down the equation you can use to work out **kinetic energy**. Include the units.

kinetic energy = ½ × _mass_ × _velocity_

(_____)   (_____) (_____)

ii Calculate the kinetic energy of Chris and her bike at 8 m/s.

kinetic energy =

kinetic energy of Chris and her bike = _____ J

iii Chris produced a driving force of 100 N. Calculate the amount of work she did.

work done = _____ × _____

work done by Chris = _____ J

iv Complete these sentences:

Chris did _____ joules of work; she gained _____ joules of _____ energy.

This is _____ than the amount of work she did to get the bicycle moving.

Remember this important principle: **Energy is always conserved**.

v In which case, explain what happened to the extra work that Chris did.

_____

_____

46

c i Complete these sentences to describe the energy change of a falling object.

When an object falls it loses _____ energy but

it gains _____ energy.

You can use this idea to work out the speed of the rollercoaster car at the bottom of the hill.

ii Write down the equation to work out how much gravitational potential energy the car loses.

_____

iii Calculate how much gravitational potential energy (gpe) the car loses.

gpe lost = _____ J

Energy is always conserved. Assume that all the gpe lost is gained as kinetic energy.

iv Write down how much kinetic energy the car has gained.

kinetic energy gained = _____ J

v kinetic energy = ½ mass × (velocity)^2

Calculate the size of the car's velocity.

car's speed = _____ m/s

This is the speed you might expect it to have.

vi In reality, how will the speed of the car compare with your answer to part v? _____

vii Explain why.

_____

Growth and development – Higher   B5

1 New cells in plants specialize into cells of roots, leaves, or flowers.

a Draw a line to match each of these key words with their meanings.

development	the sequence of growth and reproduction
growth	increase in size
life cycle	increase in complexity
specialized	adapted for a purpose

b In plants and animals, cells become specialized to do a particular job.

Use these words to complete the boxes and add another example.

organs   tissues

1 **cells:** can specialize to do a particular job

2 .... : groups of specialized cells

3 .... : groups of tissues with a particular function

For example, in plants: xylem   flower

Another example from plants is: ....

c Give an example of a specialized plant cell and explain how it is suited to do its particular job.

....

....

49

B5 Growth and development – Higher

2 Cells in an early human embryo can develop into any sort of cell, but soon cells become specialized to form a particular type of tissue.

a The diagram shows a human egg being fertilized and starting to develop.
Choose words from this list to complete the labels.

| embryo | sperm cell | nucleus | zygote | egg cell | cytoplasm | fertilization |

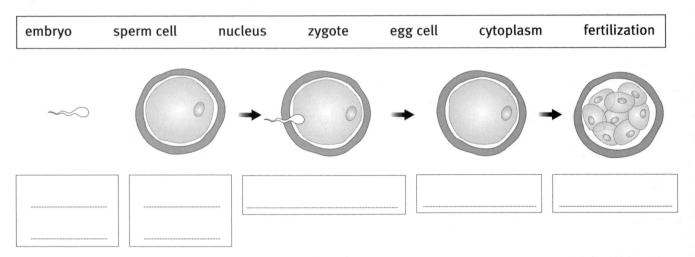

b Complete this sentence.

A human zygote must contain .... for making all the different types of cells in the human body.

c Up to the 8-cell stage, the cells in the developing embryo are not specialized.
Complete the notes below the diagram to describe an important characteristic of these embryonic stem cells.

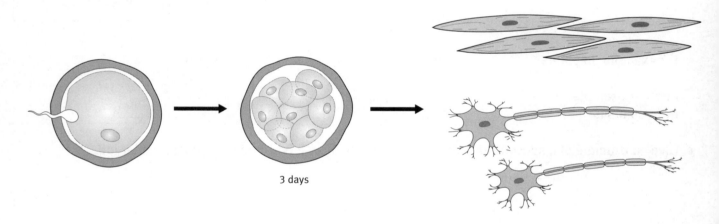

3 days

Embryonic stem cells can ....

....

Growth and development – Higher **B5**

3 Unlike most animal cells, some plant cells remain unspecialized and can develop into any type of plant cell.

a Flowering plants can continue growing all their lives. Use these words to complete the sentences.

| longer | meristem | roots | shoots | stems | taller | thicker |

Plants have unspecialized cells called _____ cells. These cells mean that plants can go on growing at the tips of the _____ and _____, and in the width of the _____. These unspecialized cells can produce different tissues so that plants can continue to grow _____, with _____ stems and _____ roots, throughout their life.

b Meristem cells mean that a piece cut off from a plant can, in the right conditions, grow into a complete new plant. Draw lines to match up these half sentences.

Plants that are genetically identical . . .	. . . clones of a plant with desirable features.
Producing plants from cuttings produces plants . . .	. . . are known as clones.
Cuttings can be used to produce . . .	. . . that are identical to the parent.
Producing plants from seeds produces plants . . .	. . . that vary, with some characteristics from each parent.

c Fill in the table to compare growth in humans and plants.

Characteristic	Humans	Plants
period of growth in size		
ability to repair slight tissue damage		
ability to regrow lost limb or branch		
ability to grow whole organism from a cutting		

4 During growth, cells divide by mitosis producing two new cells identical to each other and to the parent cell.

a Use these words to label these diagrams of two typical cells.

| chromosomes | cytoplasm | nucleus | organelles |

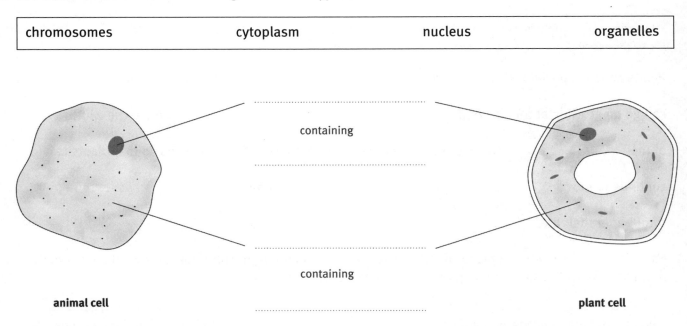

animal cell — containing — containing — **plant cell**

b Living organisms grow by making new cells. Complete the diagram to show the stages in the cell cycle in a cell with four chromosomes (in 2 pairs).

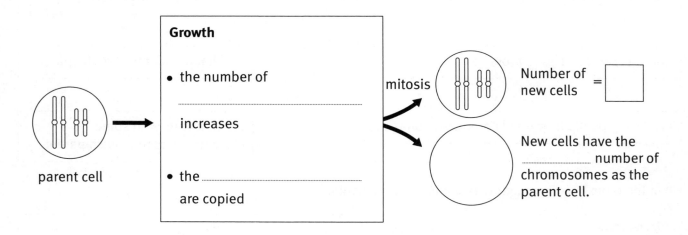

Growth
- the number of _____ increases
- the _____ are copied

parent cell → mitosis → Number of new cells = ☐

New cells have the _____ number of chromosomes as the parent cell.

c Complete this sentence.

Cell division for growth is called _____ . There are _____ new cells produced which are genetically _____ the parent cell.

5 When forming gametes (sex cells), cells divide by meiosis producing four new cells with half the number of chromosomes of the parent cell.

a Cell division that produces gametes is called meiosis. Complete the diagram showing meiosis in a cell with four chromosomes (in two pairs).

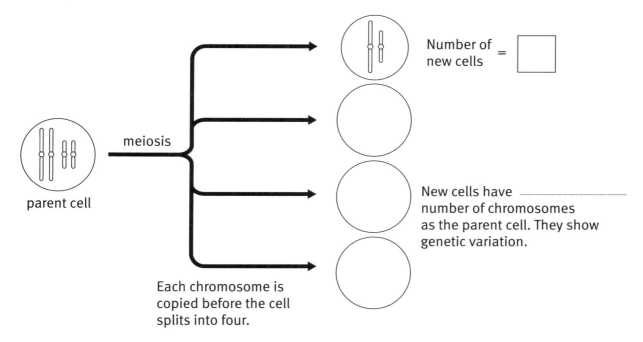

b Complete this sentence.

Cell division to form gametes is called .... There are .... new cells produced, with half the number of chromosomes of the parent cell. They show genetic ....

c In sexual reproduction, a male gamete fuses with a female gamete. The gametes each have half the number of chromosomes of the parent organism. The zygote has a set of chromosomes from each parent.

Fill in the number of chromosomes in each human gamete.

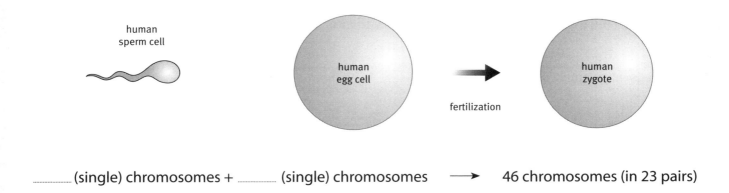

.... (single) chromosomes + ... (single) chromosomes ⟶ 46 chromosomes (in 23 pairs)

d Fill in the table (with ✓ and ✗) to compare mitosis and meiosis.

Feature	Mitosis	Meiosis
cell division for growth		
cell division for gamete production		
four new cells formed		
two new cells formed		
new cells identical to parent cell		
new cells have half the chromosomes of the parent cell		
new cells show genetic variation		
new cells genetically identical		

e Complete each of the sentences with one of these words.

| meiosis | mitosis |

- The zygote divides by _____ to form an embryo.

- Cells in the ovary divide by _____ to form egg cells.

- Cells in the root tip divide by _____ as the root grows.

- Plant cells divide by _____ to form pollen grains.

6 Genes control growth and development within the cell.

a The chromosomes in the cell nucleus are made up of many genes. Each gene is a length of DNA. DNA molecules have a double helix structure.

Use these words to complete the diagram.

| double helix | chromosomes | genes | nucleus |

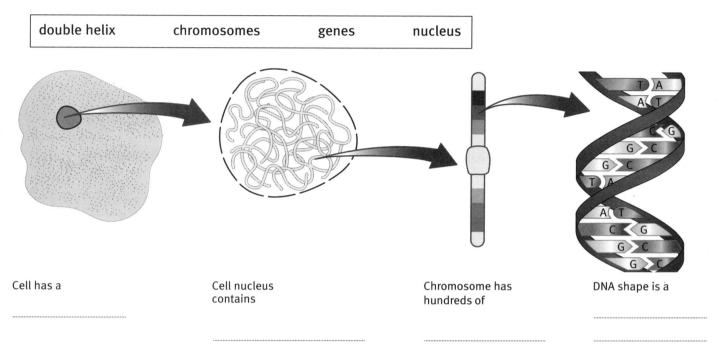

Cell has a

Cell nucleus contains

Chromosome has hundreds of

DNA shape is a

b A DNA molecule is made up of two chains connected by pairs of bases.

Draw a diagram to show the two-dimensional structure of DNA (like a ladder).

➔ Label a pair of bases (a rung).
➔ Label the two chains (the ladder sides).

7 Chromosomes are copied when the two strands of each DNA molecule separate and new strands form alongside them.

a Look at the diagram below showing how a double strand of DNA is copied to form two identical strands. Complete these sentences.

⇨ There are _____ different bases in DNA.

⇨ Base A always pairs with base _____.

⇨ Base C always pairs with base _____.

b Label the bases in the two new strands of DNA shown.

c Complete the sentences under each diagram to show how two exact copies of the DNA are made.

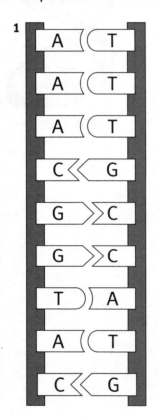

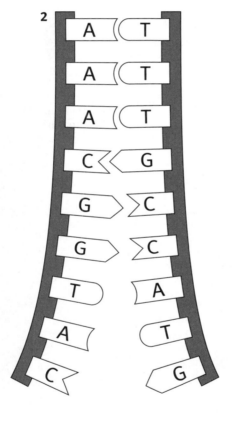

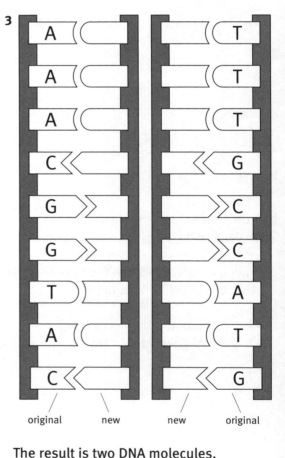

_____ along the DNA are joined in matching pairs.

Weak bonds between the bases split. The DNA opens into _____ strands. Free bases in the cell pair with the bases on each open strand.

The result is two DNA molecules. Each molecule is half new. The _____ pairs are in the same order as in the original DNA.

Growth and development – Higher **B5**

8 Body cells in an organism all contain the same genes. But many genes in a particular cell are not active because the cell produces only the proteins that it needs.

a Each gene carries the instructions for a different protein. Proteins have many different functions. Draw a line to match each protein with its description.

amylase		a structural protein of hair and nails, hard
chlorophyll		a structural protein of ligaments, strong
collagen		a structural protein of skin, stretchy
elastin		a digestive enzyme
insulin		a green pigment that absorbs light energy
keratin		a hormone involved in controlling blood sugar

b Draw a line to match each protein with the cells that make them.

amylase		hair cell
chlorophyll		salivary gland cell
keratin		green leaf cell

c Some proteins are found in every type of cell. Complete this sentence.

All cells need energy from respiration. The genes that code for enzymes used in respiration are

switched on in _____ cell.

d Cell specialization means that a cell produces only the proteins that it needs.

Complete the table with ✔ and ✗ to show which genes are active in these human cells.

Cell	Gene coding for keratin	Gene coding for amylase	Gene coding for a cell membrane protein
salivary gland cell			
hair cell			
heart muscle cell			

e Complete these sentences.

➔ Embryonic stem cells are unspecialized – all the genes are switched _____.

➔ As the cells specialize, some genes are switched _____.

➔ A cell will only make proteins from genes that are switched _____.

➔ The _____ that a cell makes control how it develops.

9 Stem cells have the potential to produce cells needed to replace damaged tissue. Genes can sometimes be reactivated in cloned cells to form cells of different tissue types.

a Complete the diagram to explain how embryonic stem cells could be used to make different types of tissue for medical treatment.

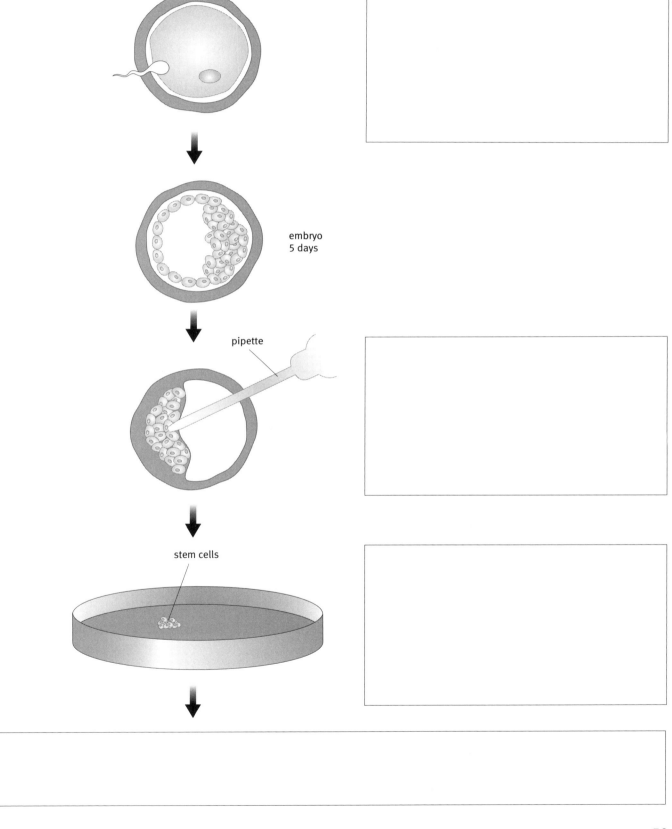

b Read the newspaper report and answer the questions.

> What does 'cloning' mean to you? Is it cloned people in scary movies? Well, don't panic. Scientists are cloning, but not to make new people. They want to find cures for illnesses.
>
> Some diseases are caused because cells in the body stop working properly. Scientists hope to clone cells from healthy people. These healthy cells would then be given to the patient.
>
> But there is a problem. Growing cells from adult humans is not very easy. Adults have only a few types of stem cells. These are unspecialized cells that can keep dividing. In adults they are used to replace worn out cells like skin and blood.
>
> Scientists are trying to find a way of 'despecializing' adult stem cells. Then adult stem cells could be used to make any type of cell.

➔ Explain how the first step in therapeutic cloning differs from the production of embryonic stem cells shown on page 13.

➔ Explain why adult stem cells are only of limited use.

➔ Explain why therapeutic cloning using an egg cell would not be necessary if scientists could find a way to 'despecialize' adult stem cells.

10 During protein production a copy of the gene is made in the nucleus. This copy carries the DNA instructions to the cytoplasm, where the protein is assembled.

a The diagram shows the steps necessary for an active gene to make a protein. Carefully label the diagram using these words.

| cytoplasm | amino acids | DNA | mRNA | nucleus | ribosome | protein |

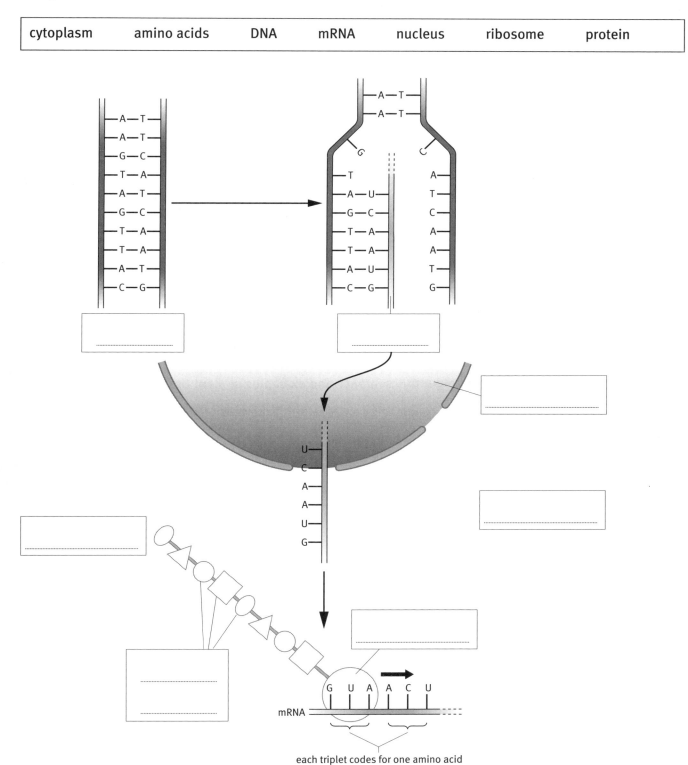

each triplet codes for one amino acid

B5 Growth and development – Higher

b Proteins are made in organelles in the cytoplasm called ribosomes. During protein production a copy of the gene is made in the nucleus (mRNA). This carries the DNA instructions to the cytoplasm, where the protein is assembled.

Put numbers in the boxes to show the sequence.

☐ In the cytoplasm, a ribosome joins amino acids together in the order coded for by the mRNA.

☐ In the nucleus, the DNA double strand separates.

☐ A molecule of mRNA forms along the DNA strand, with bases pairing with the DNA bases.

☐ The order of amino acids decides what protein is made and its particular characteristics.

☐ The small mRNA molecule passes through the pores of the nuclear membrane.

11 If the hormonal conditions in their environment are changed, unspecialized plant cells can develop into a range of other tissues or organs.

a As a plant grows, cells specialize into tissues which arrange themselves into organs.

⇨ Highlight or underline in one colour this sentence and **examples of plant tissues** in the list below.

⇨ Highlight or underline in another colour this sentence and **examples of plant organs** in the list.

⇨ Circle the **unspecialized plant cell** named in the list

| flowers | leaves | meristem | phloem | roots | xylem |

b Use these words to complete the sentences.

| tissue | roots | hormones | meristem | clone |

Unspecialized plant cells can make any kind of _____ the plant needs. Rooting powder can be used to encourage cut shoots to form _____. Rooting powder contains plant hormones.

The _____ cause the new cells produced by the _____ cells in the shoot to develop into roots. The cutting then grows into a complete plant which is a _____ of the parent.

c Explain the advantages of taking cuttings as a way of reproducing particularly good plants.

_____

_____

Growth and development – Higher **B5**

12 Phototropism increases a plant's chance of survival.

a Complete the sentence.

The rate at which a plant grows depends on the rate of photosynthesis. The rate of photosynthesis depends on the availability of carbon dioxide, water, and _____.

b Highlight or <u>underline</u> the correct description of phototropism.

- the process of using energy from light to make glucose
- the bending of plant shoots towards light
- the bending of plant shoots away from light
- providing extra light to get maximum plant growth

c Rooting powder contains plant hormones called auxins. Auxins increase the rate of plant growth. They are also involved in phototropism.

Complete the diagrams of the plants to show the way that the plant shoots would be growing.

Add notes to explain how auxins are involved.

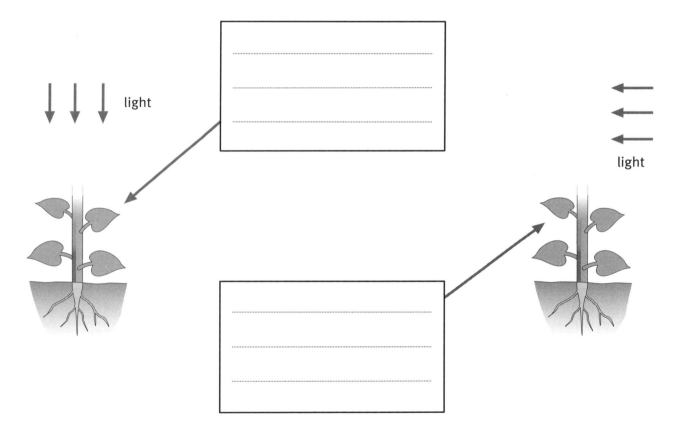

d Explain why phototropism is likely to increase a plant's chance of survival.

Chemicals of the natural environment – Higher   C5

1 Chemicals in four spheres

Write the names of these chemicals in the boxes on the diagram. Some chemicals belong in more than one box.

argon	DNA	nitrogen	sodium chloride
chalk	fat	oxygen	starch
carbon dioxide	iron ore	protein	water
crude oil	granite	sandstone	

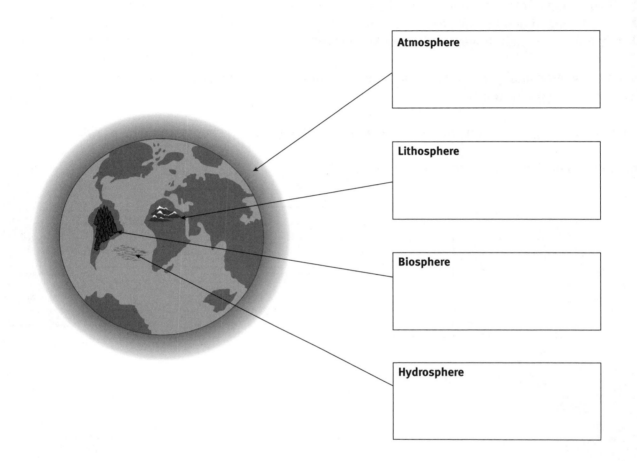

2 Gases in air

Complete this table to show the gases in unpolluted, dry air.

Gas	Element/compound	Percentage by volume in dry air
		78
	element	
argon		1
	compound	0.04

64

Chemicals of the natural environment – Higher C5

3 Chemicals of the atmosphere

Each chemical in the atmosphere consists of small molecules and is either a non-metallic element or a compound made from non-metallic elements.

a Complete this table.

Name of chemical found in air	Draw a picture of a molecule of the chemical	Write the formula that represents the molecule	Is it a non-metallic element or a compound of non-metal elements?	Is the chemical produced by human activity?
nitrogen				
oxygen				
argon				
carbon dioxide				
sulfur dioxide				
carbon monoxide				
methane				

b Complete these sentences.

⇨ Chemicals made up of small molecules have low boiling points because _____

_____

_____

⇨ Water consists of small H_2O molecules, but its boiling point is higher than molecules of comparable size and it is a liquid at normal temperatures. This is because the attractive forces between its molecules are _____ than expected.

C5 Chemicals of the natural environment – Higher

4 Strong bonds in molecules

There are strong bonds between atoms in molecules. These are covalent bonds. Colour the models and complete this table to show the bonding in molecules with the help of this information.

Atom	Usual number of covalent bonds	Colour code in models
H, hydrogen	1	white
C, carbon	4	black
O, oxygen	2	red
N, nitrogen	3	blue
Cl, chlorine	1	green

Chemical	Molecular model	Covalent bonds in the molecule	Molecular formula
hydrogen		H–H	H_2
nitrogen			
water			
methane			

Chemicals of the natural environment – Higher | C5

Chemical	Molecular model	Covalent bonds in the molecule	Molecular formula
chlorine			
hydrogen chloride			
ammonia			
ethene			

---

5 Covalent bonds

The diagram represents how a covalent bond holds two hydrogen atoms together to make an H_2 molecule.

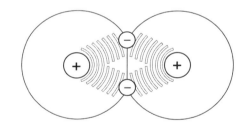

Each time a word below is printed in outline, colour it in.

Then shade the corresponding part of the diagram with the same colour.

In the hydrogen molecule (H_2) the two hydrogen atoms are held together by the electrostatic attraction between the nuclei of the two and the shared pair of electrons.

C5 Chemicals of the natural environment – Higher

6 Properties of ionic compounds

The properties of ionic compounds are explained by their structure.

a Colour the diagram of the structure of sodium chloride. Colour the sodium ions red and the chloride ions **green**. Then complete the labels.

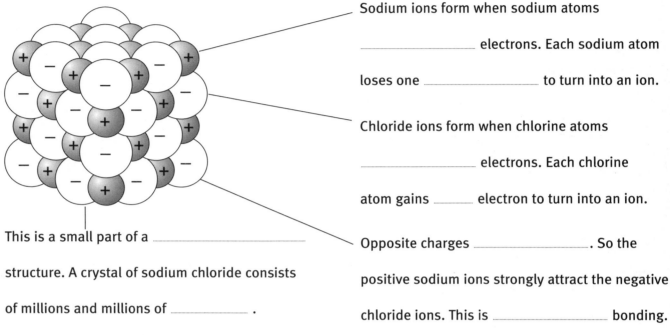

Sodium ions form when sodium atoms _____ electrons. Each sodium atom loses one _____ to turn into an ion.

Chloride ions form when chlorine atoms _____ electrons. Each chlorine atom gains _____ electron to turn into an ion.

Opposite charges _____. So the positive sodium ions strongly attract the negative chloride ions. This is _____ bonding.

This is a small part of a _____ structure. A crystal of sodium chloride consists of millions and millions of _____.

b Draw lines to match the best explanation for each property.

Property	Explanation
All the crystals of each solid ionic compound are the same shape. Whatever the size of the crystal, the angles between the faces of the crystal are always the same.	The giant ionic structure is held together by the strong attraction between the positive and negative ions. It takes a lot of energy to break down the regular arrangement of ions.
The solution of an ionic compound in water is a good conductor of electricity.	The ions in the giant ionic structure of an ionic compound are always arranged in the same regular way.
Ionic compounds have relatively high melting points.	In a molten ionic compound the positive and negative ions can move around independently.
When an ionic compound is heated above its melting point, the molten compound is a good conductor of electricity.	In a solution of an ionic compound, the positive metal ions and the negative non-metal ions can move around independently.

Chemicals of the natural environment – Higher C5

7 Silicon dioxide

Silicon dioxide is a common compound in the crust of the Earth.

a Complete the sentences beside the diagram of a quartz crystal.

Quartz is one of the crystalline forms of _____ _____.

There are small _____ crystals in granite.

The grains in sandstone consist of _____.

b Colour this diagram of the structure of quartz. Colour the oxygen atoms **red** and the silicon atoms **grey**. Then complete the labels.

This is an example of a

_____ structure.

Each oxygen atom forms

_____ covalent bonds.

Each silicon atom forms

_____ covalent bonds.

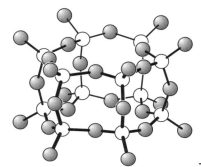

○ Si atoms   ● O atoms

The strong bonds between the atoms

are _____ bonds.

There are two oxygen atoms for every

silicon atom, so the formula is _____.

c Complete the table describing properties and uses of silicon dioxide.

Property of silicon dioxide	Comments	Use based on the property
	scratches steel	used as an abrasive in sandpaper
high melting and boiling points		used to make furnace linings and laboratory glassware
	when granite weathers, it ends up as sand in rivers and on beaches	sandstone is used as building stone
	no free electrons in the structure to carry electricity	

C5 Chemicals of the natural environment – Higher

8 The abundance of elements

Compare the percentages of elements in the Earth's crust and in the human body.

a Colour the diagrams using the same colour for any element that appears in both diagrams.

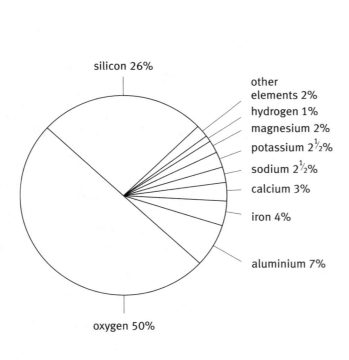
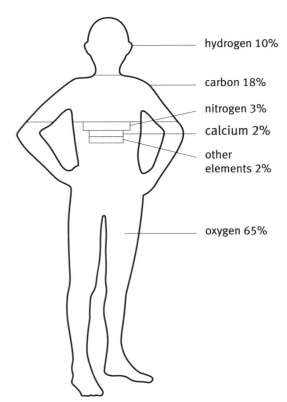

b The element which is most abundant in the Earth's crust and in the human body is ....

c Two common compounds on Earth which contain oxygen are .... and ....

d The two most abundant metals in the Earth's crust are .... and ....

e The four elements present in the largest quantities in the human body are ....

f Two 'other elements' present in the body which are important to life are .... and ....

g Molecules of carbohydrates, proteins, and DNA all include atoms of these three elements:

h The element present in DNA but not in carbohydrates and proteins is ....

Chemicals of the natural environment – Higher C5

9 Molecules in living things

Complete this table which includes information about molecules found in living things.

Molecule structure	Elements present	Formula	Type of chemical
(ribose-like sugar ring structure)			sugar
(amino acid structure with S–H side chain)			amino acid
(pyrimidine base structure)			one of the bases in DNA
(butanoic acid structure: H-C-C-C-C(=O)-O-H with Hs)			an acid found in fats

71

10 The oxygen cycle

Add to this diagram to show parts of the natural oxygen cycle. This cycle involves the element oxygen but also compounds of oxygen – including water, carbon dioxide, carbohydrates and so on.

Label the diagram to show parts of the:

| atmosphere | hydrosphere | lithosphere | biosphere |

Add arrows and labels to the diagram to show oxygen (or one of its compounds):

- moving from the biosphere to the atmosphere
- moving from the atmosphere to the biosphere
- moving from the hydrosphere to the atmosphere
- moving from the atmosphere to the hydrosphere
- moving from the atmosphere to the lithosphere

For each arrow, add a note to say what type of change is happening.

Chemicals of the natural environment – Higher   C5

11 Metals and metal ores

Complete the questions on this page with the help of the two tables.

Table 1

Metal	Metal ore	Formula of the mineral in the metal ore
aluminium	bauxite	Al_2O_3
iron	magnetite	Fe_3O_4
potassium	sylvite	KCl
tin	cassiterite	SnO_2
zinc	zincite	ZnO_2

Table 2

Reactivity	Metal
most reactive	Na
↑	Al
	Zn
The more reactive a metal is, the more strongly it holds onto oxygen and the more difficult it is to extract the metal.	Fe
	Pb
	Cu
	Ag
least reactive	Au

a Name an element from table **2** that that can be found free in nature. Why is this element found uncombined with other elements?

....

b Name three metals in table **1** that can be extracted from their oxide ores by heating with carbon.

....

c Add the missing state symbols and balance this equation that shows the use of carbon to extract a metal from an ore.

$Fe_3O_4(s) + \quad C(\_\_\_) \rightarrow \quad Fe(s) + \quad CO_2(\_\_\_)$

This equation shows the extraction of the metal iron.

The chemical that is reduced is ....

The chemical that is oxidized is ....

d Name two elements in table **1** that cannot be extracted by heating with carbon. Give a reason for your choice.

....

....

Name the method used to extract these metals ....

e Explain why it is sometimes necessary to mine large amounts of waste rock when extracting a metal such as copper.

....

....

C5 Chemicals of the natural environment – Higher

12 Calculating percentages of metal in metal ores

a Complete this diagram to work out the formula mass of the iron oxide in the ore magnetite.

Relative atomic masses: Fe = 56   O = 16

The formula

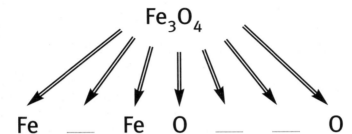

The atoms   Fe   ......   Fe   O   ....   O

The relative   ....
atomic masses

The relative formula mass of the iron oxide = ...

In this formula there are .... atoms of iron, Fe.

The relative mass of .... Fe = ...

This means that in .... kg of Fe_3O_4 there are .... kg of Fe.

So 1 kg of Fe_3O_4 contains .... kg of Fe

So 100 kg of Fe_3O_4 contains .... kg of Fe

Another way of saying this is that the percentage of Fe in Fe_3O_4 = ... %

b Work out the percentage of copper in bornite by filling in the gaps to show your working.

Relative atomic masses: Cu = 64   O = 16   S = 32

Bornite, Cu_5FeS_4, formula mass = ...

Relative mass of copper in the relative formula mass = ...

This means that in .... kg of Cu_5FeS_4 there are .... kg of Fe.

So 1 kg of Cu_5FeS_4 contains .... kg of Cu

So 100 kg of Cu_5FeS_4 contains .... kg Cu

So the percentage of Cu in Cu_5FeS_4 = ... %

13 Electrolysis of sodium chloride

a Complete the labelling of the diagram to show what happens during the electrolysis of molten sodium chloride. Choose from these words.

| carbon | chloride | electrode | ions | positive | positively | sodium |

Sodium ions are .... charged.

They move towards the negative .... during electrolysis.

Chloride .... are negatively charged. They move towards the .... electrode during electrolysis.

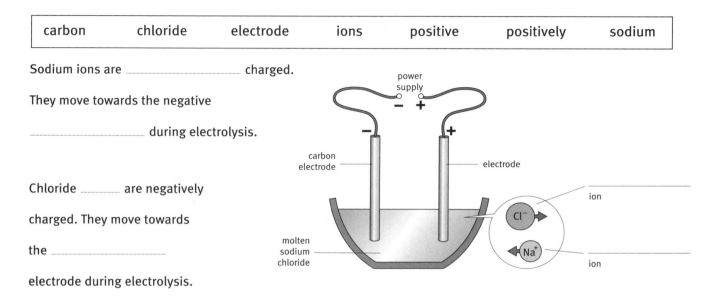

b Complete the labelling of the diagrams to explain what happens during the electrolysis of molten sodium chloride. Choose from these words.

| melts | conduct | chloride | atoms | ions | metal |
| positive | move | ions | molecules | conductor | electrons |

The .... in solid sodium chloride cannot move around, so salt does not .... electricity.

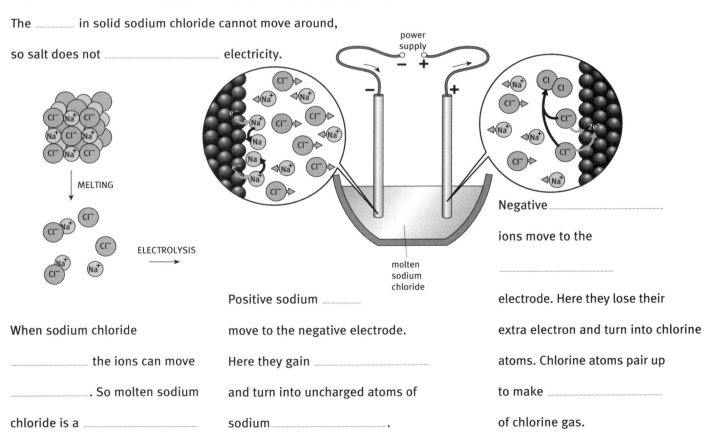

When sodium chloride .... the ions can move .... So molten sodium chloride is a ....

Positive sodium .... move to the negative electrode. Here they gain .... and turn into uncharged atoms of sodium ....

Negative .... ions move to the .... electrode. Here they lose their extra electron and turn into chlorine atoms. Chlorine atoms pair up to make .... of chlorine gas.

14 Electrolysis of aluminium oxide

a Label the diagram to describe the equipment used to extract aluminium from aluminium oxide. Use these words and phrases.

| carbon anodes | carbon lining | negative electrode |
| molten aluminium oxide | molten aluminium | tapping hole |

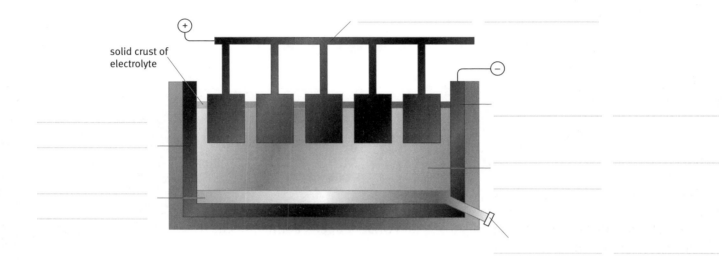

b Explain why aluminium oxide conducts electricity only when liquid but not when solid.

c Write the symbols for the two ions in aluminium oxide: _____ _____

d Complete this equation to show what happens at the negative electrode during the electrolysis of molten aluminium oxide.

_____ + 3e⁻ → Al

e Complete these equations to show what happens at the positive electrode during the electrolysis of molten aluminium oxide.

_____ → O + 2e⁻

_____ + _____ → O$_2$

15 Properties of metals

a Read the following paragraph and underline four different properties of metals.

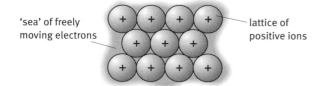

A model of metallic bonding

In a metal, such as copper, the atoms are packed closely together. The atoms are held together by strong metallic bonds, so copper is strong and difficult to melt. Copper is malleable, which means that it can be beaten into a different shape because the atoms can be moved around without the structure losing its strength. When a metal structure is formed, the metal atoms lose their outer electrons and form positive ions. The electrons are no longer held by particular atoms, so they can move freely between the positive ions. This is why metals are good conductors of electricity.

b List the four properties you underlined above in this table. In the second column, make notes on how metallic bonding explains each property.

Property	Explanation
1	
2	
3	
4	

16 The life cycle of a metal

Use these words and phrases to complete the diagram which shows the life cycle of a metal. Two of the boxes have been filled in for you.

metal in use	rubbish to waste tip	mining the ore
separating and purifying the mineral with the metal	separating and recycling waste metal	recycling scrap metal
making products from the metal	extracting the metal from the mineral	shaping the metal into sheets, wire, or bars
end of useful life		

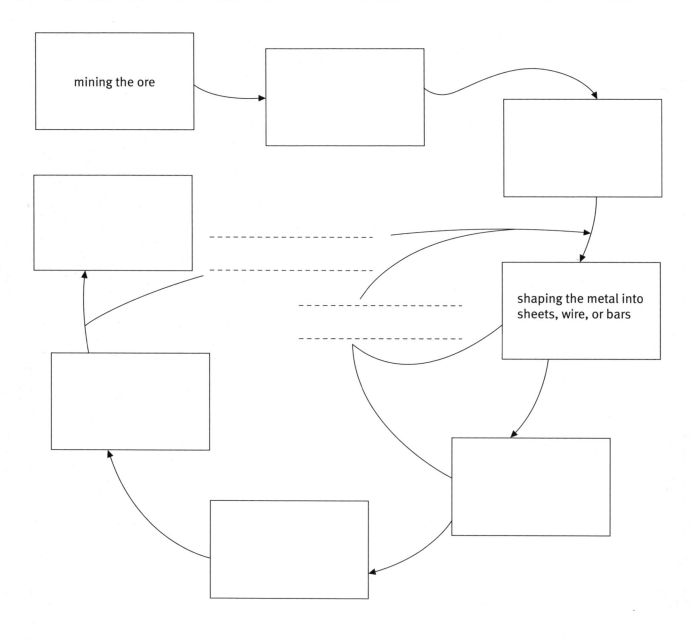

Electric circuits – Higher   P5

1 Static

Insulating materials get charged up when they are rubbed. This is due to stationary charge. It is called static electricity.

a A piece of polythene is rubbed with a cloth and gets charged. Another piece of polythene is rubbed with the cloth.

  i Complete these sentences. Draw a ring round the correct **bold** words.

  The two pieces of polythene will get **the same / different** charge. This means that they will **attract / repel** each other. The polythene was charged because of the movement of **electrons / ions** which have a **positive / negative** charge.

  ii Look at the picture below. It shows that the polythene became negatively charged. There is a line that shows the transfer of electrons. Draw an arrow on one end of this line to show the direction in which the electrons moved.

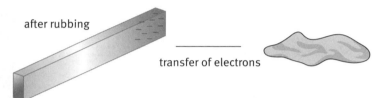

  iii Explain how the neutral cloth has become positively charged. _____

  _____

  iv Draw a line to match the start of each sentence with its correct ending. These sentences summarize the forces between different charges.

  | Like charges . . . | . . . attract each other. |
  | Opposite charges . . . | . . . repel each other. |

b Polythene gets a negative charge when it is rubbed. Perspex gets a positive charge.

  i Complete these sentences. Draw a ring round the correct **bold** words.

  A piece of perspex will **attract / repel** another piece of perspex. A piece of perspex will **attract / repel** a piece of polythene.

  ii Polythene repels nylon. What charge must nylon have? _____

  iii What effect will Perspex have on nylon? _____

c The dome of a Van de Graaff generator gets charged up when it is switched on. Yasmin is holding the dome and her hair stands on end.

  i Explain why her hair stands on end.

  _____

  ii The teacher discharges the dome and Yasmin's hair falls back down. Rearrange the word below to describe Yasmin's charge once her hair has fallen back down.

  unrealt   _____

P5 Electric circuits – Higher

2 Current in a series circuit

a Physicists think that an electric current is the flow of charge – the same charge that causes static electricity. The statements below explain one piece of evidence for this belief. But they are out of sequence. Use arrows to join up the statements in the correct order. The first one has been done for you.

| This implies that the electric current, |

| When the spark jumps across the gap, |
↓

| These are the same charges that jump across the gap to make a spark. |

| is a movement of charges. |

| which makes the lamp light up, |

| the lamp lights up. |

b The picture shows a model of an electric circuit. Bethan is the 'battery'. She pushes the loop of rope around in the direction shown by the arrows. The other pupils let it pass through their hands.

Join the boxes below to show how the model helps to explain an electrical circuit. The first linking line has been done for you.

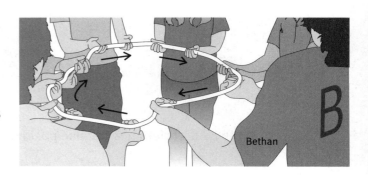
Bethan

Rope circuit

| When Bethan first pushes the rope, it starts moving through everyone's hands at the same time. |

| Bethan gets tired after pushing the rope around. |

| The others feel their hands getting hot. |

| If any one of the others grips the rope firmly, the rope stops moving. |

| At any time, the amount of rope leaving each child's hand is the same as the amount going in. |

Electrical circuit

| Stored energy is transferred out of the battery. |

| Putting in an insulator stops the flow of charge. |

| The current is not used up. It is the same everywhere. |

| Charge moves throughout the circuit as soon as it is connected up. |

| The battery does work on all other components in the circuit. |

Electric circuits – Higher   P5

c All matter is made of atoms, which contain protons (positive) and electrons (negative). In metals, the electrons are free to move throughout the metal. That is why metals are electrical conductors.

The flow of charge around a circuit is called electric current. Traditionally, physicists and engineers use 'conventional current', showing the direction that positive charges would be flowing. The electrons actually flow in the other direction.

i Look at the descriptions below. Draw lines to match current with its description in the middle and its direction on the right.

| 'conventional current' | flow of negative electrons | from − terminal to + terminal |

| actual movement | imaginary flow of positive charges | from + terminal to − terminal |

ii The diagram shows atoms in a wire.

⇨ Label the end of the wire joined to the positive (+) battery connection.

⇨ Draw an arrow next to the wire to show the direction of 'conventional current'.

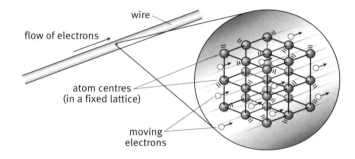

d i This diagram shows a computer model of an electric circuit.

Complete the sentences.

The electric current is being measured by

an ....

The meter reading is 10 milliamps (mA). This is the same as

.... amps (A).

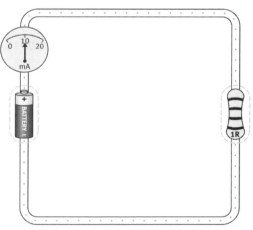

ii This diagram shows the same circuit, but it now has two extra meters.

Put arrows in each meter to show the readings, then complete the sentences.

The charges in an electric circuit never get used up.

Current is the *flow* of charges, so the current is

.... all the way round the circuit.

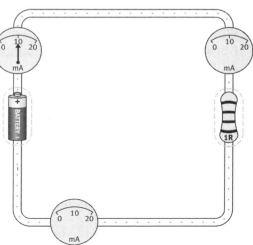

3 Current in parallel circuits

When several components are connected in parallel directly to a battery

X → the current through each component is the same as if it were the only component present

Y → the total current from (and back to) the battery is the sum of the currents through each of the parallel components

Circuit 1 on the right shows a single resistor connected to a battery.

The current through the resistor is 6 mA.

a What is the current coming from the battery?

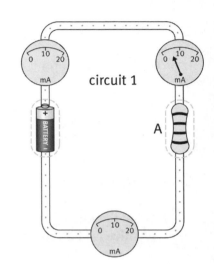

circuit 1

b Put arrows in the other two ammeters to show their readings.

In circuit 2, another similar resistor has been added in parallel.

c What current will the current be in resistor A in circuit 2?

d Which of the statements at the top of the page tells you this? Draw a ring round the correct statement below.

statement X   statement Y

e Draw arrows in ammeters 1 and 2 to show their readings.

f Resistor B is exactly the same as resistor A. What will the current be in resistor B?

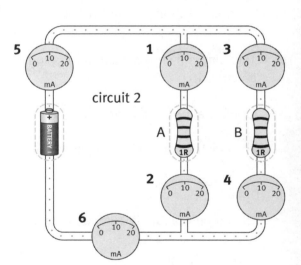

circuit 2

g Draw arrows in ammeters 3 and 4 to show their readings.

h What will be the current through the battery?

_____ mA

i Which of the statements at the top of the page tells you this? Draw a ring round the correct statement below.

statement X   statement Y

j Draw an arrow in each of the ammeters 5 and 6 to show their readings.

k Now imagine you add a third resistor in parallel. Put a tick (✓) next to the true statement below.

→ The current through the third resistor will be the same as the currents through resistors A and B and the current coming from the battery will increase.

→ The current coming from the battery will stay the same and the current in resistors A and B will drop slightly.

4 Resistance

a Look at the statements on the left below. Some of them apply to electric current and some apply to electrical resistance.

Draw a line to link each statement with the correct box on the right. The first one has been done for you.

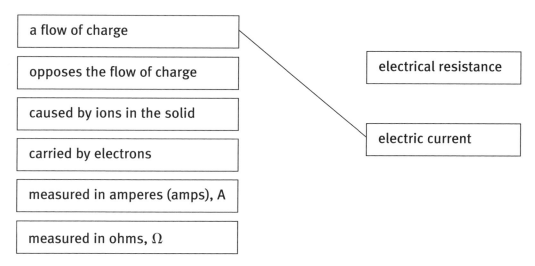

- a flow of charge
- opposes the flow of charge
- caused by ions in the solid
- carried by electrons
- measured in amperes (amps), A
- measured in ohms, Ω

- electrical resistance
- electric current

b Complete the sentences below. Draw a ring round the correct **bold** words.

➔ The resistance of a circuit determines the **size** / **direction** of the current.

➔ A big resistance makes it **more** / **less** difficult for the charge to flow and leads to a **big** / **small** current.

➔ It is easier for charge to flow through a **smaller** / **bigger** resistor so the current through it is **smaller** / **bigger**.

➔ Insulators have a very **small** / **big** resistance and so the current through them is practically zero.

c Combining resistors in series and parallel will make a new resistance.
  i Complete the statements below (left). Draw a ring round the correct **bold** words.

  ii Draw a line to link each statement with the correct explanation on the right.

Two resistors in series have a **bigger** / **smaller** resistance than either one on its own.	Because there are more paths that the moving charges can take.
Two resistors in parallel have a **bigger** / **smaller** resistance than either one on its own.	Because the moving charges have to pass through one then the other.

d Look at the networks of resistors below. Each resistor has the same value. Put them in order of increasing resistance. Number them from 1 to 5: the smallest has the number 1 and the largest is 5. The first one has been done for you.

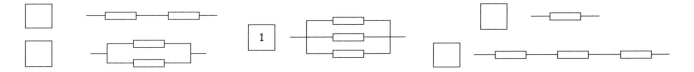

P5 Electric circuits – Higher

5 Ohm's law

a Use these words to complete the sentences below. Words can be used once, more than once, or not at all.

| voltage | size | bigger | smaller | double | battery | proportional |

A .... pushes charge around an electric circuit. A battery's .... is a measure of its push. The bigger a battery's voltage, the .... its push on the charge. In turn, this will lead to a .... current. The current is .... to the voltage. This means that doubling the voltage will .... the current.

b Look at the circuit on the right. It has two 1.5 V batteries giving a total voltage of 3 V. The current in the circuit is 24 mA.

Now look at the circuits below. The resistance of each circuit is the same. However, the number of 1.5 V batteries changes.

   i In each circuit, write the voltage across the resistor next to the voltmeter

   ii In each circuit, write the current next to the ammeter.

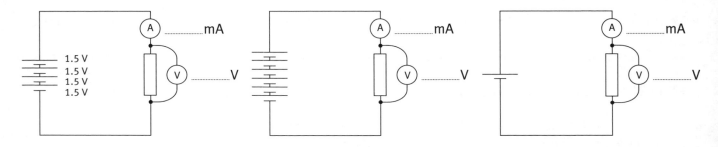

c The graph on the right shows how the current varies in a fixed resistor as the voltage is increased.

   i Complete this sentence by filling in the missing word.

   The current is .... to the voltage.

   ii Imagine the test is done on a resistor with twice the resistance.

   What difference will this make to the currents at each voltage?

   ....

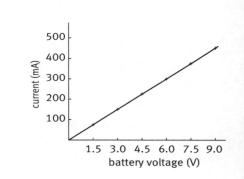

   iii Draw in the line you would get for current against voltage with twice the resistance.

d The equation below is used to calculate resistance. Complete the equation by putting in the units.

$$\text{Resistance (............, }\Omega\text{)} = \frac{\text{voltage (............, V)}}{\text{current (............, A)}}$$

Electric circuits – Higher   **P5**

6 Resistance of LDR and thermistor

a i Some electrical components have a resistance that changes. Look at the three components below left.

⇨ Draw a line from each component to match it to the description of how to change its resistance.

⇨ Draw a line to join each of the descriptions to a possible use on the right.

The first one has been done for you.

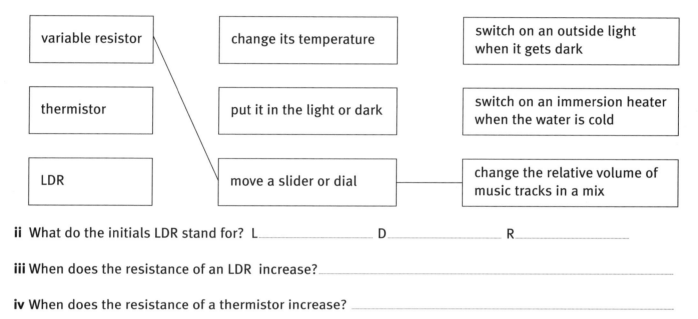

ii What do the initials LDR stand for? L................ D................ R................

iii When does the resistance of an LDR increase? ....

iv When does the resistance of a thermistor increase? ....

b Use the names in the list below to label the circuit symbols.

| single battery | power supply | filament lamp | switch | LDR | resistor |
| variable resistor | thermistor | ammeter | voltmeter | | |

—(A)— ....   —(V)— ....   —(⊗)— ....

—[ ]— ....   —[⁄]— ....   —[⁄]— ....

—o/ o— ....

—|⊢— ....   —o o— ....   —(⊘)— ....

85

7 Potential difference

The voltage of a battery is the 'push' the battery gives the charge. But it is also the work the battery does in pushing each unit of charge around the circuit. Inside the battery, the chemical reactions give each charge some **potential energy**.

The water model can help to show this. The pump pushes on the water, raising it up and increasing its potential energy. It loses this potential energy as it flows round the circuit.

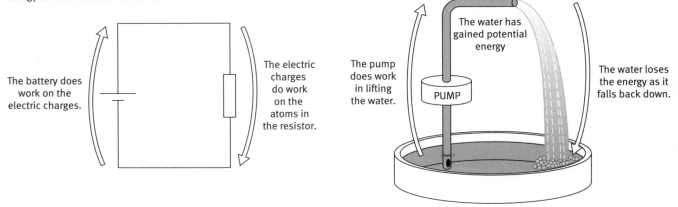

a Join the boxes below to show how the model helps to explain the energy changes in an electrical circuit. The first one has been done for you.

| The battery pushes on the charge, raising its potential energy. | The water loses potential energy as it falls into the tray. |

| The resistor heats up as the moving charge does work on its atoms. | The pump pushes on the water, raising it up to where it has more potential energy. |

| The charge loses potential energy as it does work in the resistor. | The water heats up slightly when it splashes into the tray. |

| The potential energy lost by the charge in the resistor is the same as the potential energy it gained in the battery. | The water loses the same potential energy when it falls into the tray as it gained in the pump. |

b As charge flows around a circuit, its potential energy changes. The charge gains potential energy in a battery and loses potential energy in resistors (or other components).

Fill in the blanks to complete the sentences below.

→ The p_____ d_____ between two points in a circuit is measured in v_____.

→ Voltage is another word for p_____ d_____.

→ The bigger the p_____ d_____ between two points in a circuit, the more energy is transferred between these points.

c Look at the circuits on the right. The batteries and lamps are identical. The only difference is that circuit B has two of the batteries in parallel.

Look at the comparisons in the table below. Put a tick (✓) in the column that correctly describes how they compare in the two circuits.

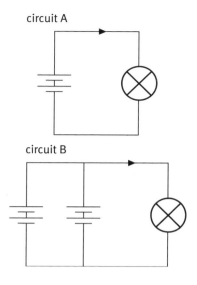

circuit A

circuit B

	the same	bigger in A	bigger in B
the potential difference across the bulb			
the current through the bulb			
the current through a single battery			
the time for the batteries to go flat			

d Potential differences add up around a circuit.

Look at the circuit on the right and the statements below. Draw a line to match each of the statements on the left with its explanation on the right. The first one has been done for you.

The bigger the potential difference across the battery, the bigger its push.	Energy is conserved in the circuit: there is no net gain or loss of energy.
There is a drop in potential across each of the resistors.	The harder the battery pushes the charge, the more work it does.
There is a bigger potential difference across the bigger resistor.	The charge does work as it moves through a resistor.
The sum of potential differences across two resistors equals the potential difference across the battery.	The charge does more work as it goes through a bigger resistor.

P5 Electric circuits – Higher

8 Current and p.d. in circuits (series circuits)

a Complete the sentence below. Draw a ring round the correct bold words.
When two resistors are in series

- the current through the bigger resistor will be **bigger than / the same as / smaller than** the current in the smaller resistor.
- The potential difference across the bigger resistor will be **bigger than / the same as / smaller than** the potential difference across the smaller resistor.

b Look at the circuits below.

  i For each circuit, put a tick (✓) next to the resistor which has the bigger potential difference across it.

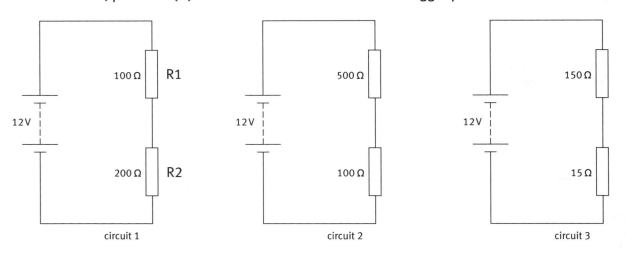

circuit 1   circuit 2   circuit 3

  ii In circuit 1, the total resistance is 300 Ω.

   Calculate the current in the circuit. Use the equation current = voltage/resistance.

   current I = $\dfrac{V}{R}$ = $\dfrac{\text{_____ V}}{\text{_____ Ω}}$ = _____ A

  iii Calculate the potential difference across each of the resistors in circuit 1.
   Use the equation voltage = current × resistance.

   p.d. across across R1 = _____ A × _____ Ω = _____ V

   p.d. across across R2 = _____ A × _____ Ω = _____ V

  iv Does this agree with the tick you put on circuit 1 in part **i**? _____

  v Explain why the sum of the potential differences in part **iii** must be 12 V.

   ....

  vi Draw in a voltmeter in circuit 3 to show how you would measure the voltage across the 150 Ω resistor.

9 Current and p.d. in circuits (parallel circuits)

a Complete the statements below. Draw a ring round the correct **bold** words.
When two components are in parallel

- the potential differences (voltage) across each one are **the same / different**
- the potential difference across each one is **equal to / smaller than** the voltage of the battery

b A hair dryer has a heater and a fan motor in parallel. There are two switches. When switch **A** is closed, the fan comes on. When switch **B** is closed, the heater comes on as well. The heater will not come on unless switch **A** is already closed.

i Complete the circuit diagram below right by labelling the two switches **A** and **B**.

ii Suggest why it is important that the hair dryer cannot be turned on with the heater on and the fan off.

....

....

....

iii The motor has a resistance of 460 Ω and the heater has a resistance of 46 Ω. The mains supply is 230 V. Calculate

- the current in the 46 Ω heater

$$\text{current } I \quad \frac{-}{-} = \frac{V}{R} = \frac{\text{........ V}}{\text{........ Ω}} = \text{........ A}$$

- the current in the 460 Ω motor

$$I = \frac{V}{R} = \text{........ A}$$

- the total current from the 230 V supply = ... A

iv Complete these sentences. Draw a ring round the correct **bold** words.

When two components are in parallel, the current in each one is **the same as / smaller than** the current for that component on its own. The total **current / voltage** taken from the power supply will be **more / less** than it was with just one component on its own. This extra current will have a cost: the time for the batteries to go flat will **increase / stay the same / decrease**.

v Look at the list of words below. Draw a line under those that go *through* a component and a ring round those that go *across* a component.

| potential difference | flow | current | voltage |

10 Power, fuses and cables

a In electric circuits, the equation for power is

    power   =   potential difference (voltage)   ×   current

  (..........................)   (..........................)   (..........................)

i Complete the equation above by putting in the units.

In symbols this is $P = VI$.

ii Rearrange the equation so that the current, I, is the subject:

$I = $ ☐

iii A 60 W bulb is connected to the mains. The mains voltage is 230 V. What current will it take? (You may find a calculator useful here.)

$I = $ ☐ = ... A

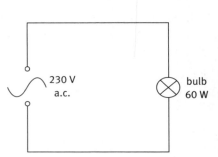

b Power measures the rate at which energy is transferred. Complete the sentences below. Draw a ring round the correct **bold** words.

A 100 W light bulb is **brighter / dimmer** than a 60 W bulb. It transfers energy **more / less** quickly. They are both connected to the same, mains voltage of **230 V / 12 V**. This means that **more / less** charge has to flow through the 100 W bulb each second. So the current in the 100 W bulb is **bigger / smaller**.

c In a house, light bulbs are connected in parallel. The picture on the right shows a circuit with three 100 W bulbs. Each bulb takes a current of 0.44 A.

i What will the current be at the point marked X? .... A

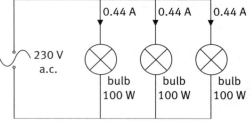

ii The cable in the circuit has a current rating of 5 A. To be safe, the current in these cables must be less than 5 A. How many 100 W bulbs can be put in parallel before the current in the cable becomes unsafe at the point marked X?

....

iii Why is it dangerous for the current in the cable to go over 5 A? ....

....

iv A 5A fuse is put in a house lighting circuit to protect the cables. On the circuit above, mark the place to put the fuse.

v Explain how the fuse protects the cables in this circuit. ....

....

11 Domestic appliances and bills

a Power measures how fast we transfer energy, so

energy transferred = power × time

i complete these sentences.

⇨ The equation 'energy = power × time' gives the energy in joules when power is measured in

_____ and time is measured in _____

⇨ A joule is a small unit for everyday purposes, so when we pay electricity bills the unit of energy is the

kilowatt hour (kWh).

⇨ To calculate the energy in kWh, we use the same equation 'energy = power × time' but we measure

power in _____ and time in _____ .

ii Asif puts a 60 W light on for 3 hours.

⇨ Write 60 W in kilowatts: _____ kW
⇨ How many units (kWh) does the light bulb use in that time?

Energy = power × time

= _____ kW × _____ hour = _____ kWh

iii Electricity supply companies charge about 10 p for each kWh. How much does it cost to have the light on?

cost = 10 p per kWh × _____ kWh = _____ p

b Filament light bulbs are not very efficient. They produce light by getting very hot. The diagram below shows how the energy supplied to a light bulb by electricity is then transferred away from the bulb.

i How does the diagram show that only 25% of the energy supplied is usefully transferred?

_____

_____

_____

ii Name a kind of light bulb that is more efficient.

_____

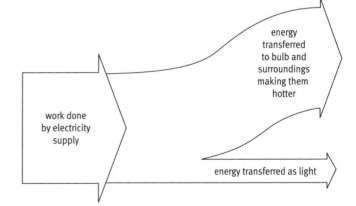

c The equation for efficiency is:

$$\text{efficiency} = \frac{\text{useful energy transferred}}{\text{total energy supplied}} \times 100$$

An electric saw does 150 J of useful work for every 400 J supplied by the mains. What is its efficiency?

Efficiency = _____ = _____ %

12 Generator effect

a A moving magnet induces a voltage in a coil of wire.
The picture shows the north pole of a magnet moving into a coil.
The needle on the ammeter flicks to the right.

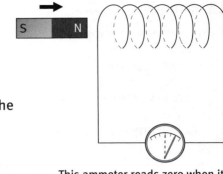

This ammeter reads zero when its needle is in the middle.

Look at the boxes below. Draw a line to match each of the actions on the left with one of the movements of the needle on the right.
You can use each needle movement once, more than once, or not at all.

| Pull the north pole out of the coil |
| Hold the magnet stationary in the coil |
| Push the south pole into the coil |
| Pull the south pole out of the coil |

| flick to the left |
| no movement |
| flicks to the right |

b The magnet can be put on a spindle and rotated near the coil. This will induce an alternating current in the coil.

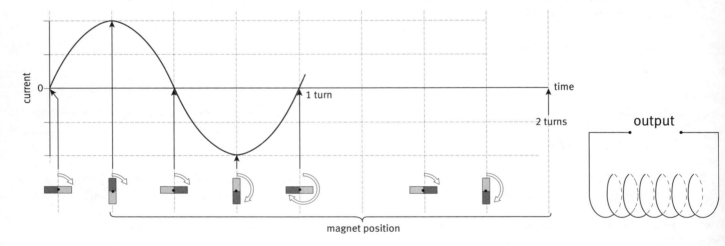

This diagram shows how the induced current varies as the magnet rotates through different positions. It shows one cycle of the a.c.

i Complete the diagram, showing how the current varies through two full rotations of the coil.

ii Draw in the missing magnet positions.

iii Complete the sentences.

In the generators of UK power stations, the time for one rotation is $\frac{1}{50}$ of a second. This means that

the frequency (number of cycles in one second) is .... hertz (Hz).

iv Write down four ways in which you could induce a bigger voltage in the coil.

....

....

13 Transformers

a Look at the picture of a simple transformer.

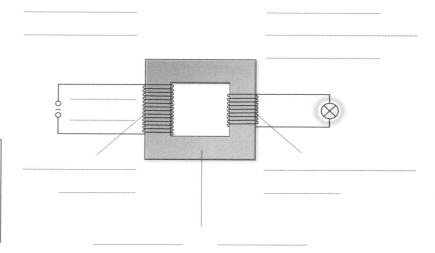

i Complete the diagram using these labels.

soft iron core	primary coil
secondary coil	a.c. supply
induced alternating current	

ii Complete these sentences. Draw a ring round the correct **bold** words.

A transformer works because the current in the **primary / secondary** coil produces a **magnetic / electric** field which passes through the secondary coil. This field is changing and therefore **induces / reduces** a voltage.

The transformer above has **more / fewer** turns on the secondary coil. This means it is a **step up / step down** transformer. The output voltage will be **more / less** than the input voltage.

iii This is the transformer equation: $\dfrac{V_p}{V_s} = \dfrac{N_p}{N_s}$

The input voltage is 230 V, and the number of turns are 460 (primary) and 24 (secondary). What is the output voltage from the transformer?

$V_s = $ _____ V

iv The National Grid distributes mains electricity from power stations to people's homes.

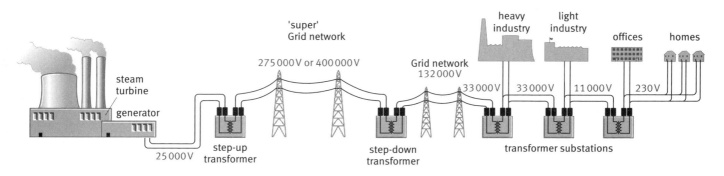

Look at the phrases below. Some of them apply to a.c. and some to d.c. Put a <u>straight line</u> under the phrases that refer to d.c. and a <u>curvy line</u> under the phrases that refer to a.c.

It is easier to generate.   Its voltage has a constant value.   It can be distributed more efficiently.

It is produced by batteries.   It comes from the mains supply.   It won't pass through a transformer.

This page is blank

Brain and mind – Higher B6

1 Animals respond to stimuli in order to keep themselves in favourable conditions.

a A stimulus is a change in the environment of an organism. Look at the examples of behaviour below.

⇨ Highlight or underline in one colour this sentence and **the stimuli**.

⇨ Highlight or underline in another colour this sentence and **the responses**.

> Woodlice prefer dark places; they move away from light.
>
> Bacteria living in the gut move towards the highest concentration of food.
>
> An earthworm rapidly withdraws into its burrow if pecked.
>
> A resting housefly flies takes off as soon as it sees any fast movement nearby.
>
> If an octopus sees a predator, it releases a cloud of 'ink' and moves away quickly.

b Simple reflexes are automatic. They give responses that help an animal survive and reproduce. Complete this list.

Simple reflex behaviour helps an animal to:

⇨ find _____

⇨ escape from _____

⇨ find a _____

⇨ avoid _____ environments

c Simple animals rely on reflex actions for most of their behaviour. This means they cannot adapt their behaviour, or learn from experience. Explain the disadvantages of this.

_____

_____

_____

d Fill in the words that best fit these descriptions. You will find all the words you need on this page.

⇨ an action or behaviour caused by a stimulus   _____

⇨ a change in the environment that causes a response   _____

⇨ the way an organism responds to a stimulus or situation   _____

⇨ a response to a stimulus that is involuntary   _____

B6 Brain and mind – Higher

2 Simple reflexes produce rapid involuntary responses.

a Simple reflexes are responses that you do not think about or learn. Look at the list below. Highlight or underline the examples of simple reflexes.

- Your pupils get smaller in bright light.
- You answer a question.
- Your eyes water on a windy day.
- A newborn baby grasps at anything put in her hand.
- A goalkeeper saves a goal.
- You breathe faster when you run.

b Humans and other mammals have very complex behaviour, but simple reflexes are also important for their survival. Complete the descriptions of human reflexes in the tables, and add some more examples.

Adult reflex	Stimulus	Response
gagging	touch to back of the throat	closing of throat
pupil		

Newborn reflex	Stimulus	Response
grasping	touch to palm of hand	
stepping		walking movement of legs
sucking		

c Many newborn reflexes are present for only a short time after birth. Explain why they increase a young baby's chances of survival. Use these words in your explanation.

behaviour experience learn

Brain and mind – Higher **B6**

3 Some complex organs have receptors and effectors.

a Multicellular organisms respond to stimuli through **receptor** and **effector** cells.
Fill in the gaps using the bold words.

detection of stimulus → response to stimulus

_____ cell → _____ cell

b Different receptors detect different types of stimuli. List some examples of:

→ receptors that detect changes outside the body _____

_____

→ receptors that detect changes inside the body _____

_____

c Some receptors are made up of single cells. Others are grouped together as part of a complex organ. Complete the sentences.

Single-cell receptors are found in human _____, for example _____ receptors.

An example of a human sense organ is the _____, which detects _____.

d Responses to stimuli are carried out by effector organs. Effector organs are either glands or muscles. Use these words to complete the sentences.

| contract | hormones | move | secreting |

→ The effector cells in glands respond to stimuli by _____ chemicals, for example

_____, enzymes, or sweat.

→ The effector cells in muscles respond to stimuli by causing the muscle to _____ and

_____ a part of the body.

4 Responses are coordinated by the central nervous system (CNS). Sensory and motor neurons carry the signals.

a In mammals the nervous system is made up of a **central nervous system** (the **brain** and **spinal cord**) connected to the body via the **peripheral nervous system**.

Label or colour the diagram to show the parts of the human nervous system printed in bold.

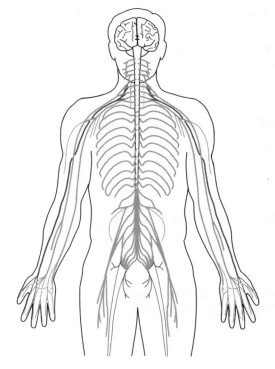

b Neurons are cells in the nervous system that carry nerve impulses. Use these words to label the diagram of a neuron.

| axon | cell membrane | cytoplasm | fatty sheath | nucleus |

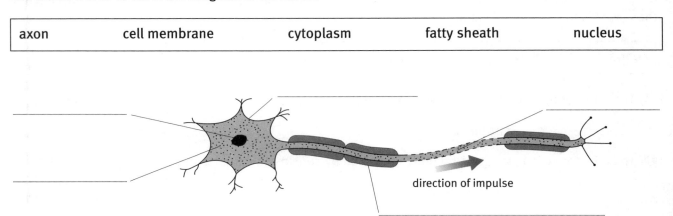

direction of impulse

c Neurons transmit electrical impulses when stimulated. Use these words to complete the sentences.

| axon | electrical | insulates | speed |

When the neuron is stimulated, an _____ impulse travels along the _____ to the branched ending. Here it connects with another neuron or an effector. Some axons are surrounded by a fatty sheath, which _____ them from neighbouring cells and increases the _____ of the nerve impulse.

d **Sensory** neurons carry impulses from **receptors** to the CNS. **Motor** neurons carry impulses from the CNS to **effectors**.

Label the diagram using the bold words.

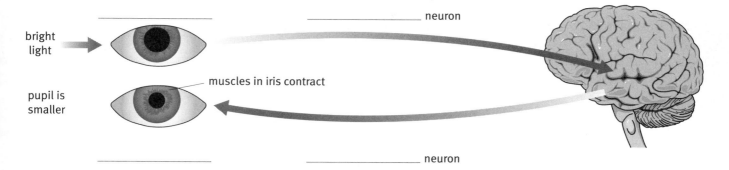

e The diagram above shows a reflex arc coordinated by the brain. Reflexes are involuntary actions, and most are coordinated by the spinal cord.

Fill in the flow diagram to describe the route of the nerve impulses when you pick up a hot plate. (The diagram above will help you.)

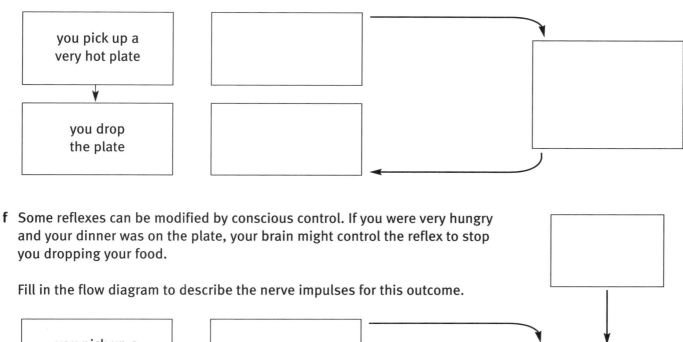

f Some reflexes can be modified by conscious control. If you were very hungry and your dinner was on the plate, your brain might control the reflex to stop you dropping your food.

Fill in the flow diagram to describe the nerve impulses for this outcome.

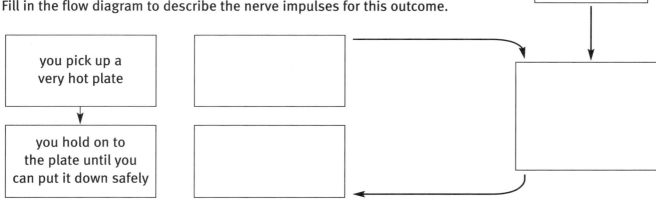

5 Chemicals released into the synapses transmit nerve impulses from one neuron to the next.

a Synapses are tiny gaps between neurons. Electrical impulses cannot jump across synapses. Chemicals carry impulses between neurons.

Use the diagrams of an impulse crossing a synapse to put the events below in order. Number the boxes 1 to 5.

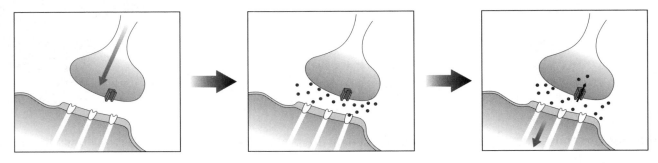

☐ The chemical is absorbed back into the sensory neuron, to be used another time.

☐ An impulse is stimulated in the motor neuron.

☐ A nerve impulse travels along a sensory neuron until it reaches a synapse.

☐ The molecules diffuse across the synapse. They bind to receptor molecules on the membrane of the motor neuron.

☐ The end of the sensory neuron releases a chemical into the synapse.

b The receptor molecules only bind to certain chemicals. Complete the diagrams.

⇨ Add receptors of the correct shape to synapse **A**.

⇨ Add chemicals carrying the impulse to synapse **B**.

⇨ Add arrows to show the direction of the nerve impulse across these synapses.

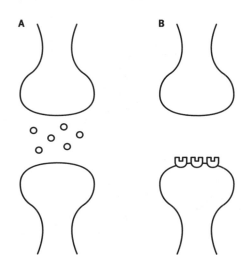

6 Some drugs and toxins affect the way impulses cross synapses.

a Use colours or lines to match up these words with their descriptions.

caffeine	a poison that causes dangerous effects in the body
curare	a medicine or other substance that causes effects in the body
drug	a stimulant present in tea, coffee, and soft drinks
Ecstasy	a painkiller used as a medicine
morphine	a substance that increases nervous activity
painkiller	a substance that reduces the sensation of pain
stimulant	a poison that blocks transmission of nerve impulses (and stops breathing)
toxin	a drug that has mood-enhancing effects

b The substances described above can affect the transmission of nerve impulses.

  i What substance(s) from the list above might reduce or stop the transmission of impulses at synapses?

  ....

  ii What substance(s) from the list above might increase the transmission of impulses at synapses?

  ....

  iii Alcohol is another drug that affects the transmission of nerve impulses at synapses. It has a different effect in low and high doses. Suggest what each effect is on the nerve synapses in the brain.

  ➔ Low dose: ....

  ➔ High dose: ....

c Highlight or colour the drug that would interfere with the transmission of an impulse in the synapse shown in the diagram.

d Explain how this drug would stop the nerve impulse from crossing the synapse.

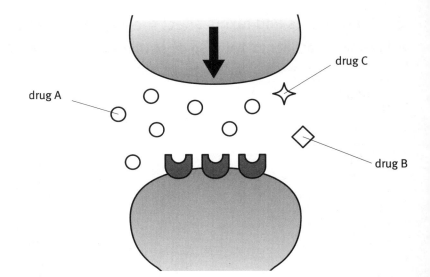

_____

_____

_____

e Some drugs work by blocking the reuptake of the synapse chemical. Serotonin is released at one type of synapse in the brain. It triggers nerve impulses causing feelings of pleasure. The drug Ecstasy (or MDMA) blocks the reuptake of serotonin.

Look at the diagram and answer the questions.

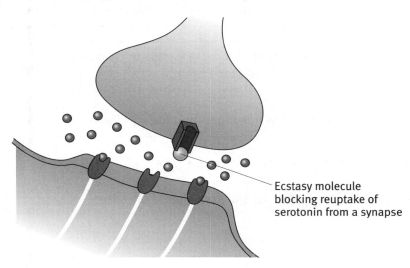

Ecstasy molecule blocking reuptake of serotonin from a synapse

i What effect does the presence of the Ecstasy molecule have on the amount of serotonin in the synapse?

_____

ii What effect does this have on the receptor molecules detecting serotonin?

_____

iii What effect does this have on the activity of this neuron?

_____

f Complete the sentence.
The mood-enhancing effects of Ecstasy are due to the _____ in serotonin concentration at synapses in the brain.

7 The cerebral cortex is the part of the brain most concerned with intelligence, memory, language, and consciousness.

a Choose one of these words to complete the sentence.

| hundreds | thousands | millions | billions |

The human brain is made of _____ of neurons.

b Draw a line to match this word to the best description.

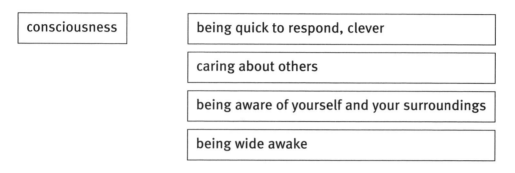

c The diagram shows the right side of a human brain that has been cut in half.

→ Colour the part most involved in intelligence, memory, language, and consciousness.

→ Then complete the description.

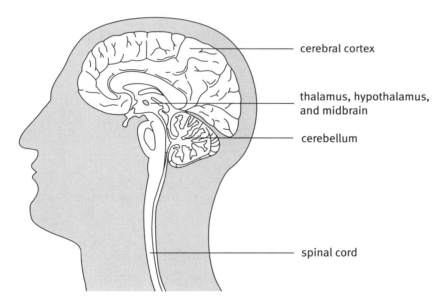

The _____ is highly folded, giving it a _____ surface area.

Different areas are responsible for different _____. It is much bigger in _____ than in other mammals.

d Studies of patients whose brains have been partly destroyed by injury or disease tell us about the functions of different areas of the cortex.

Use the information below to decide which area of the brain has been damaged in stroke patients with the symptoms shown in the table.

Symptom	Damaged brain region (A to E)
speech is slurred	
has trouble controlling hand movements	
legs feel numb	

Part of cerebral cortex	Function
A speech centre	talking
B sensory cortex	receiving information from receptors
C motor cortex	voluntary movement
D visual cortex	detecting visual stimuli
E language	understanding language

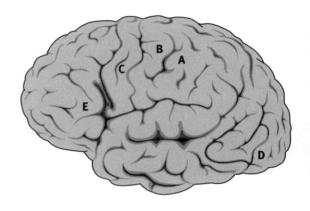

e Regions of the cortex have also been mapped using electrical stimulation during open brain surgery. Explain how this can show the function of different areas of the motor cortex.

....

....

....

f Modern MRI brain scans give information about the function of different parts of the cortex. Explain the advantages of using MRI.

....

....

....

8 Conditioned reflexes can be learned. The final response has no direct connection to the stimulus.

a Animals can learn to link a new stimulus with a reflex action. This is a conditioned reflex response. Pavlov showed this in experiments with dogs. Read what he did and answer the questions.

Pavlov's dog salivated when presented with food.

Pavlov rang a bell while his dog was eating its food.

After a while the dog salivated when it heard the bell, even if no food was around.

  i What was the primary stimulus (that originally caused salivation)? ....

  ii What was the reflex response? ....

  iii What was the secondary stimulus (that Pavlov added)? ....

  iv What was the conditioned reflex response? ....

b Complete the sentences.

Salivation is a .... response connected to food. Salivation and hearing a bell have no direct connection. In a .... reflex, the response has no direct connection to the stimulus.

c Conditioned reflexes can increase chances of survival. Read the text in the box and answer the questions.

> Many birds feed on caterpillars. Some caterpillars are brightly coloured and taste nasty (to the birds).
> Young birds try eating the caterpillars and learn that they taste nasty. In future they avoid brightly coloured caterpillars.

  i What was the primary stimulus? ....

  ii What was the secondary stimulus? ....

  iii Which animal had an increased chance of survival? ....

  iv Explain how hoverflies increase their chances of survival by having markings that make them look like wasps.

B6 Brain and mind – Higher

9 Learning is the result of experience. It creates pathways in the brain that are more likely to transmit impulses than others.

a Draw a line to match each of these key words with their meanings.

adapting	knowledge or skills gained from experience
learning	doing the same thing more than once
neuron pathways	adjusting to new conditions
repetition	strong links between points along connecting neurons

b Human babies learn very quickly. Number these sentences to explain the sequence of events in the brain during learning. One has been done for you.

1	The cortex in a baby's brain is a complicated network of neurons.
	The response is learned.
	Strengthened connections make it easier for more impulses to travel along the pathway.
	A new experience sets up new pathways between the neurons in the cortex.
	Using the pathway strengthens the connections between the neurons.

c Explain why repetition helps you learn a new sporting or musical skill.

....

....

....

Brain and mind – Higher B6

10 An animal can adapt well to new situations if it has a variety of potential pathways in the brain.

a Adapting to new situations means learning new skills. Explain why humans are good at learning new skills throughout their life.

_____

_____

_____

b As we get older it becomes harder for the language-processing area in the cortex to make new pathways. Explain how this would affect an adult learning a new language.

_____

_____

c Describe how you could strengthen new pathways in the language-processing area of the cortex when learning a new language.

_____

d Use these words to complete the sentences.

| destroyed | development | difficulty | experience |

As the brain develops, _____ strengthens some neural pathways. Some connections that haven't been strengthened by experience are _____. There is evidence from studying child _____ that children may acquire some skills only at a particular age. Studies of rare cases of 'feral' or neglected children, who have not learnt any language in early childhood, show that they have _____ learning language skills later in their childhood.

B6 Brain and mind – Higher

11 Memory means stroing and retrieving information.

a Highlight or underline two phrases that together describe memory.

| learned behaviour | storage of information | processing of information |
| retrieval of information | input of sensory information | |

b Complete the sentence.

Memory concerning words or language is called _____ memory.

c Memory can be divided into short-term memory and long-term memory. Complete the table comparing these.

Memory type	How long does it last?	How much can be stored?	An example
short-term			remembering that this row is about short-term memory
long-term			remembering the date of your birthday

d Short-term and long-term memory work separately in the brain. Describe some evidence for this.

_____

_____

_____

12 Scientists have models of how memory works, but they do not explain memory completely.

a Explain, in terms of neuron pathways in the brain, why repetition will help you remember something.

b Explain, in terms of short-term and long-term memory stores, why repetition will help you remember something.

c Look at this list of numbers for 20 seconds. Cover the list, then try to write the sequence down.

| 1 | 2 | 3 | 5 | 8 | 13 | 21 | 34 | 55 | 89 | 144 |

d If you found it difficult, look for a pattern in the numbers then try again.

e Some memories are triggered by particular sounds, sights, or smells. Describe, with an example, how a strong stimulus can make you more likely to remember something.

f Give an example of a method you use to help you remember important information. Then try to explain why it works by using a memory store model.

Chemical synthesis – Higher   C6

1 The chemical industry

This chart shows the value of the sales of the various sectors of the chemical industry in the EU. The figures are for 2004.

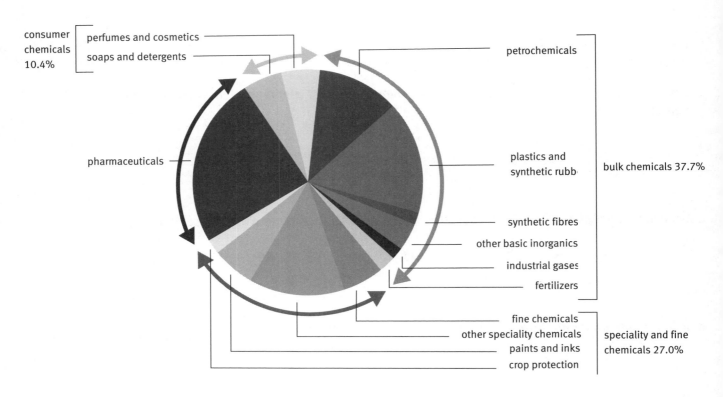

a What is the raw material for making petrochemicals? ....

b Give an example of a petrochemical that is used to make plastics on a large scale. ....

c Give an example of a type of chemical needed for 'crop protection'. ....

d Why are paints and inks now so valuable as products? ....

e Give the chemical formulae of the molecules of these three gases that are used in industry:

- nitrogen ....
- oxygen ....
- hydrogen ....

f What percentage value of products of the EU chemical industry are used:

- as perfumes, cosmetics, soaps, and detergents? ....
- for diagnosis and treatment in medicine? ....

Chemical synthesis – Higher C6

2 Acids and alkalis

a Complete this table about some acids.

Name of acidic compound	Formula of acidic compound	State of the pure compound at room temperature
citric acid	$C_6H_8O_7$	
tartaric acid	$C_4H_6O_6$	
	H_2SO_4	
	HNO_3	
ethanoic acid	CH_3COOH	
hydrogen chloride		

b Complete this table about some alkalis.

Name of alkaline compound	Formula of alkaline compound	State of the pure compound at room temperature
	NaOH	
potassium hydroxide		
calcium hydroxide		solid

c Acids and alkalis show their characteristic reactions when dissolved in water. Draw a line to match each of these solutions to its pH value.

pure water	pH 14
dilute hydrochloric acid	pH 12
dilute sodium hydroxide solution	pH 7
vinegar	pH 3
limewater (calcium hydroxide solution)	pH 1

C6 Chemical synthesis – Higher

3 Reactions of acids

- Complete the general word equations to summarize the main reactions of solutions of acids in water.
- Next complete the word equations for the examples.
- Finally complete and balance the matching symbol equations, including state symbols.

a Acids with metals

The general word equation for the reaction of an acid with a metal is:

acid + _____ → salt + hydrogen

Example

_____ + magnesium → magnesium chloride + hydrogen

HCl(___) + _____ → MgCl$_2$(___) + _____(___)

b Acids with metal oxides

The general word equation for the reaction of an acid with a metal oxide is:

acid + metal oxide → salt + _____

Example

nitric acid + copper oxide → _____ + _____

_____(___) + CuO(___) → Cu(NO$_3$)$_2$(___) + _____(___)

c Acids with metal hydroxides

The general word equation for the reaction of an acid with a metal hydroxide is:

acid + metal hydroxide → _____ + water

Example

_____ + _____ → sodium sulfate + _____

H$_2$SO$_4$(___) + NaOH(___) → _____(___) + H$_2$O(___)

d Acids with metal carbonates

The general word equation for the reaction of an acid with a metal carbonate is:

acid + metal carbonate → _____ + carbon dioxide + water

Example

_____ + _____ → calcium chloride + carbon dioxide + _____

HCl(___) + CaCO$_3$(___) → _____(___) + _____(___) + H$_2$O(___)

4 Ions and formulae

The table shows the charges on common ions found in ionic compounds.

a Complete the table

Positive ions			Negative ions		
Ion	Charge	Symbol	Ion	Charge	Symbol
lithium	....	Li$^+$	chloride	1–	....
sodium	1+	....	....	1–	Br$^-$
....	1+	K$^+$	iodide	....	I$^-$
magnesium	2+	....	nitrate	1–	NO$_3^-$
....	....	Ca^{2+}	....	1–	OH$^-$
barium	2+	Ba^{2+}	oxide	2–	....
....	3+	Al^{3+}	....	....	CO$_3^{2-}$
			sulfate	....	SO$_4^{2-}$

b Use the complete table to help you to write down the formulae of these ionic compounds.

sodium hydroxide .... magnesium carbonate ....

sodium chloride .... magnesium sulfate ....

sodium nitrate .... calcium carbonate ....

sodium carbonate .... calcium chloride ....

potassium chloride .... calcium iodide ....

magnesium bromide .... calcium nitrate ....

magnesium oxide .... aluminium oxide ....

magnesium hydroxide .... aluminium chloride ....

c Work out the symbol for the ions in these compounds that are not included in the table.

→ sulfide ion in barium sulfide, BaS ....

→ strontium ion in strontium nitrate, Sr(NO$_3$)$_2$ ....

→ phosphate ion in sodium phosphate, Na$_3$PO$_4$ ....

C6 Chemical synthesis – Higher

5 Testing the purity of citric acid

Procedure
(1) A 0.48 g sample of citric acid was dissolved in 50 cm³ of water.
(2) 1 cm³ phenolphthalein indicator was added.
(3) The solution was titrated with a solution of sodium hydroxide. These were the burette readings from the titration:

	Titration
Second burette reading/cm³	22.40
First burette reading/cm³	3.30
Volume of NaOH(aq) added/cm³	

a Label these diagrams to summarize the procedure.

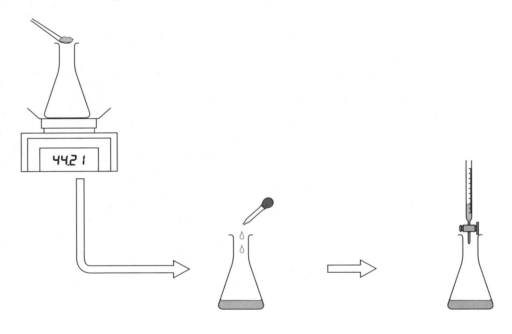

b Calculate the purity of the citric acid from the formula given below.
   T is the titre (the volume of solution added from the burette).
   The concentrations of the sodium hydroxide used in the titration meant that the value of $F = 0.025$.

$$\text{purity} = \frac{T \times F \times 100}{\text{mass of sample}} \%$$

purity = _____ %

c Give an example to show why it is important to be able to measure the purity of a chemical such as citric acid.

6 Effect of surface area on the rate of a reaction

The diagrams show the apparatus for investigating the rate of reaction of marble chips with acid.
→ Diagram **A** shows the apparatus before the reaction starts.
→ Diagram **B** shows the reaction in progress.

The reaction was carried out twice – first with larger chips of marble, then with smaller chips. There were still unchanged marble chips in both flasks when the reaction stopped.

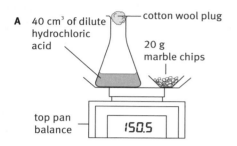

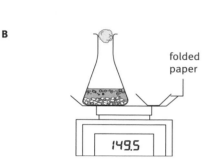

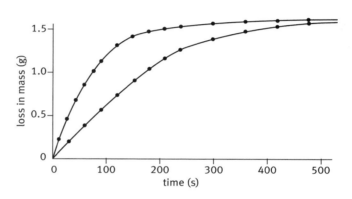

The graph shows typical results using two sizes of marble chips.

a Label the graph lines to show:
→ which is the line for larger chips and which the line for smaller chips
→ where the reaction was fastest
→ where the reaction was slowing down
→ where the reaction had stopped.

b Explain why the total mass of the flask, acid and marble fell during the reaction.

c Explain why the reaction slowed down and stopped with the same final loss in mass for both the larger chips and the smaller chips.

d Explain the difference in the rate of reaction at the start for the two sizes of marble chips.

7 Effect of concentration on the rate of reaction

Adding dilute hydrochloric acid to a solution of sodium thiosulfate starts a slow reaction.

$$Na_2S_2O_3(aq) + 2HCl(aq) \rightarrow 2NaCl(aq) + H_2O(l) + SO_2(aq) + S(s)$$

The mixture turns cloudy. In time it is not possible to see through the solution.

The diagram shows a method for investigating the rate of reaction. The experimenter looks down at the cross on the paper from above and records the time it takes for the cross to vanish after adding the acid.

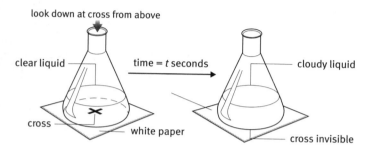

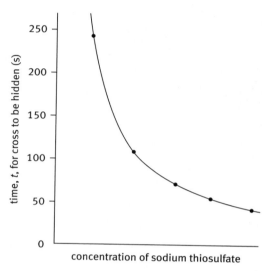

In an investigation, an experimenter added 5 cm³ of dilute hydrochloric acid to 50 cm³ samples of sodium thiosulfate solution. The experimenter measured the time for the cross to disappear with five different concentrations of sodium thiosulfate solution. The graph is a plot of the results.

a Look at the equation above and use it to explain why the solution of sodium thiosulfate turned cloudy after the hydrochloric acid was added.

b Why did the experimenter use the same volume and concentration of dilute hydrochloric acid with each different concentration of sodium thiosulfate solution?

c Put into words what the graph shows about the effect of changing the concentration of the sodium thiosulfate solution on the rate of the reaction.

Chemical synthesis – Higher C6

8 The effect of temperature changes on the rate of reaction

The reaction of sodium thiosulfate with hydrochloric acid (see question 7) can also be used to investigate the effect of temperature on the rate of a reaction. In this investigation, the volume and concentration of the acid and sodium thiosulfate stay the same. The thiosulfate solution is warmed before adding the acid. The experimenter measures the temperature of the mixture after adding the acid and the time taken for the cross to disappear.

Temperature (°C)	20	30	40	50	60
Time taken for the cross to disappear (s)	280	132	59	31	17

Explain what you can conclude from the results in the table.

....

....

....

9 Predicting rates of reaction

The graph shows the volume of hydrogen produced when excess zinc granules react with 50 cm³ dilute hydrochloric acid at 20 °C, plotted against time.

Show the effect of each change to the conditions by completing this table and also by drawing on the graph the line you would expect.

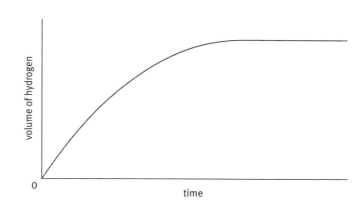

Change of conditions (all other factors stay the same)	Effect on the rate of reaction at the start	Effect on the volume of gas collected (at room temperature) when the reaction stops
halving the concentration of the acid		
carrying out the reaction at 30 °C		
using the same mass of zinc but in larger pieces		

117

10 The effect of catalysts on rates of reaction

Hydrogen peroxide solution contains the compound H_2O_2.

At room temperature it decomposes very slowly to give water and oxygen.

$2H_2O_2(aq) \rightarrow 2H_2O(l) + O_2(g)$

The graph shows the results from three tests. Each time a small amount of a metal oxide was added to 50 cm³ of a solution of hydrogen peroxide in a flask. The oxygen gas was collected and measured for up to 5 minutes.

In a control experiment, with no added metal oxide, no oxygen was collected in 5 minutes.

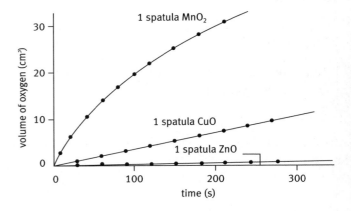

a Complete and label this diagram to show how the oxygen could be collected and measured.

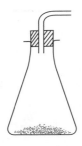

b Put the three metal oxides tested in order of their effectiveness as catalysts for the reaction.

....

c Why was a control experiment carried out?

....

d Explain the meaning of the term 'catalyst'.

....

....

....

11 Collision theory

Chemists use collision theory to explain why factors such as surface area, concentration, temperature, and catalysts affect the rates of reactions.

a Write a paragraph of four or five sentences to explain the key ideas of collision theory.

b Colour and label this diagram. Then use it in an explanation to show how collision theory explains why reactions in solution go faster if the concentrations of reactants are higher.

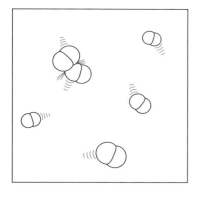

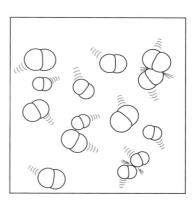

Explanation:

12 Making a soluble salt from an acid

The diagrams show how to make a pure sample of magnesium sulfate from magnesium oxide.

Label the diagrams. Add a caption to each stage of the diagram to describe what is happening and to explain the purpose of the stage.

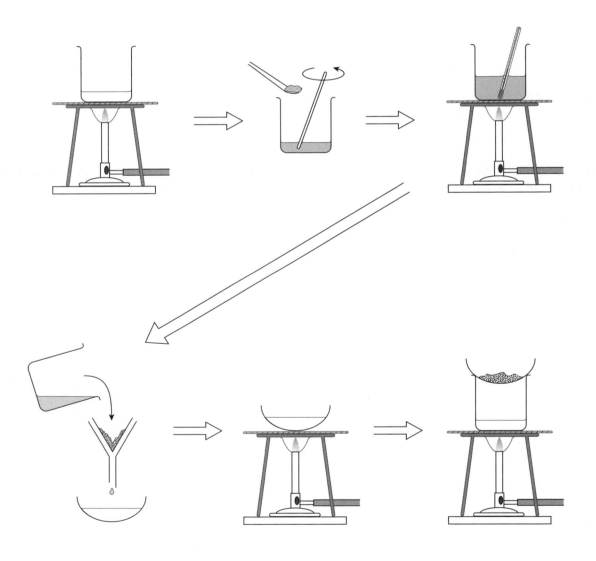

Chemical synthesis – Higher — C6

13 Stages in synthesis

Use the example on page 28 (or any other example) to explain the importance of these stages in any chemical synthesis.

a Choosing the reaction

b Working out the quantities to use

c Carrying out the reaction in suitable apparatus under the right conditions

d Separating the product from the reaction mixture

e Purifying the product

f Measuring the yield and checking the purity of the product

C6 Chemical synthesis – Higher

14 Yields

a Work out the theoretical yield of magnesium sulfate, $MgSO_4$, that can be made from 4.0 g of magnesium oxide, MgO, and an excess of sulfuric acid.

→ **Step 1** Write the balanced symbol equation for the reaction. (Write it on the top line only.)

→ **Step 2** Work out the formula masses of the chemicals.

Relative atomic masses: Mg = 24, O = 16, S = 32

→ **Step 3** Write the relative reacting masses for the relevant chemicals under the balanced equation in step 1.

→ **Step 4** Convert to reacting masses by adding the units.

→ **Step 5** Scale the quantities to the amounts actually used to find the theoretical yield.

b What is the percentage yield if the actual yield is 10.0 g?

c Use the same steps to work out the theoretical yield of zinc sulfate, $ZnSO_4$, that can be made from 9.3 g zinc carbonate, $ZnCO_3$, and an excess of sulfuric acid. (Relative atomic masses: Zn = 65, C = 12, S = 32, O = 16.)

d What is the percentage yield if the actual yield is 11.4 g?

15 Neutralization

a Draw a line to match each statement on the left with the related box on the right.

The salt formed when sodium hydroxide reacts with sulfuric acid	NaOH
The salt formed when potassium hydroxide reacts with citric acid	soluble metal hydroxide
The acid which reacts with calcium hydroxide to form calcium nitrate	neutralization
The alkali which reacts with acetic acid to form sodium acetate	sodium sulfate
A set of compounds that are alkalis in water	HNO_3
The type of reaction that occurs when an acid and an alkali form a salt	potassium citrate

b The graph shows how the pH of a solution of hydrochloric acid changed with the volume of dilute sodium hydroxide added.

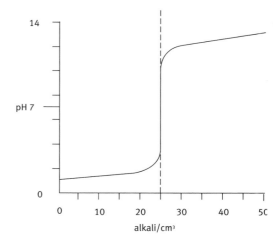

→ What volume of the dilute sodium hydroxide solution was needed to neutralize the dilute acid?

Explain your answer.

....

....

....

→ The neutralization can be carried out with a few drops of universal indicator in the solution. Use colours to indicate on the graph the colours you would expect to see at each stage as the acid is added to the alkali.

....

....

....

....

C6 Chemical synthesis – Higher

16 Ionic theory of neutralization

a Use these words and symbols to complete the text.

acid	alkali	hydrogen	hydroxide	ions	molecules	salt	sodium
$H^+(aq)$	$H_2O(l)$	$H_2O(l)$	$K^+(aq)$	$NO_3^-(aq)$	$OH^-(aq)$		

- Acids are chemicals containing _____ which react with water to give hydrogen _____ in solution.

   $HNO_3(l) + water \rightarrow$ _____ $+ NO_3^-(aq)$

- Alkalis are ionic compounds. Examples are the soluble hydroxides of the alkali metals (lithium, _____, and potassium). These compounds consist of metal ions and _____ ions. When they dissolve they add hydroxide ions to water.

   $KOH(aq) \rightarrow K^+(aq) +$ _____

- Potassium hydroxide and nitric acid react to make a _____ (potassium nitrate) and water.

   _____ $+ OH^-(aq) + H^+(aq) +$ _____ $\rightarrow K^+(aq) + NO_3^-(aq) +$ _____

- During a neutralization reaction the hydrogen ions from the _____ react with the hydroxide ions from the _____ to make water _____.

   $H^+(aq) + OH^-(aq) \rightarrow$ _____

   The remaining ions in solution make a salt.

b The diagram illustrates the changes to ions when dilute hydrochloric acid reacts with dilute sodium hydroxide solution. Complete the diagram.

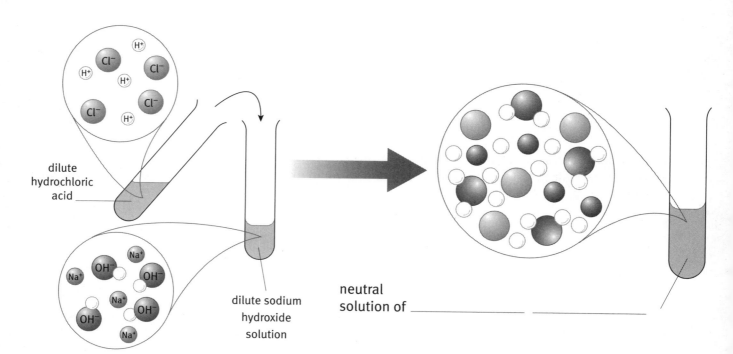

The wave model of radiation – Higher   P6

1 Vibrations make waves

a The table shows some examples of waves.

In each case:

- underline the medium that carries the wave (the first one has been done for you)
- suggest what might have started the wave

Wave	What might have started the wave?
a sound wave in the <u>air</u>	
a ripple on the water of a pond	
a pulse on a slinky	
a wave on a stretched elastic rope	

b Complete these sentences. Draw a (ring) round the correct **bold** words.

The coils of a slinky spring are **disturbed / increased** by a wave. However, once the **disturbance / dismay**

has passed, the coils return to **where they started / the end of the spring**.

c Look at the quantities below. Draw a (ring) round those that can be permanently displaced by a wave – the wave carries them from one place to another.

| energy | matter | particles | information |

d i Use these words to complete the boxes on the left.

transverse   **longitudinal**

_____ waves – when the particles of the medium vibrate at right angles to the direction in which the wave is moving

_____ waves – when the particles of the medium vibrate in the same direction as the wave is moving

The pictures on the right show the two different types of wave on a slinky spring. The waves are moving left to right.

ii Draw arrows on each wave to show the direction in which the coils are moving.

iii Draw lines to link each box on the left with one of the waves on the right.

P6 The wave model of radiation – Higher

e Use these words to complete the sentences.

| second | pitch | vibrating | hertz |

Sound is produced by something _____. The frequency of the vibrations determines the _____ of the sound. The unit of frequency is the _____ (Hz), which measures the number of vibrations every _____.

f i Draw a line to match each note on the left with the type of vibration that produces it.

Note	Type of vibration	Screen display of waves
High pitch note	Vibration with a low frequency	(wave with fewer cycles)
Low pitch (deep) note	Vibration with a high frequency	(wave with more cycles)

The sound waves can be picked up by a microphone and displayed on a screen. These are shown on the right.

ii Draw a line to match each type of vibration with one of the screen displays on the right.

g Cross out the wrong definitions to complete this table for a transverse wave.

amplitude	The maximum distance each particle moves from its normal position	or	The total distance from the top of a crest to the bottom of a trough
frequency	The time it takes a wave to pass a point	or	The number of waves passing a point every second
wavespeed	The speed at which the wave moves up and down	or	The speed at which the wave moves through the medium
wavelength	The length of a complete wave, e.g. from one crest to the next	or	The total distance that a wave has travelled from its source

h Mark on this diagram of a transverse wave:
- the **amplitude**
- the **wavelength**

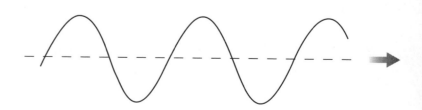

2 How waves move

The frequency of a wave depends on the source – how many times it vibrates per second. The wavespeed depends on the medium the wave travels through.

a Wavespeed can be calculated using this equation:

      wavespeed   =   frequency   ×   wavelength

    (..)   (..........................)   (..........................)

  i Complete the equation by putting in the units.

  ii Rewrite the equation using these symbols: λ f v   .... = ... × ....

  iii What is the symbol for wavelength? ....

b The diagram shows a burst of three waves on a rope.

The frequency is 3 Hz and their wavelength is 4 m.

Write your answers on the diagram.

  i How many complete waves pass a point in 1 second?

  ii What is the wavelength of each wave?

  iii What is the total length of this burst of three waves?

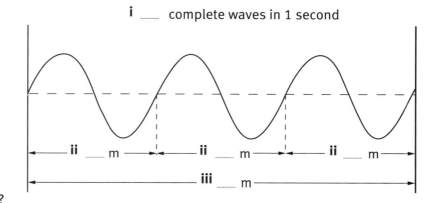

These three waves pass a point in 1 second.

  iv Complete this equation to find the speed of the wave:

    wavespeed = 3Hz × 4 m = ... m/s

c Another wave has a wavelength of 2 m and a frequency of 6 Hz. Use the equation to find the speed of this wave.

  wavespeed = ... × .... = ... m/s

d The waves in parts **b** and **c** were made on the same piece of rope.

Complete these sentences. Draw a (ring) round the correct **bold** words.

The speeds of the waves in parts **b** and **c** are **the same / different**. This is usually the case for waves in the same **medium / direction**: their speed **does / does not** change even if they have different frequencies.

In summary: usually the speed of a wave is **independent of / proportional to** its frequency.

P6 The wave model of radiation – Higher

3 Water waves

a The diagrams below show four ways in which water waves behave. Use the words in the list to label the diagrams.

| refraction | reflection | interference | diffraction |

i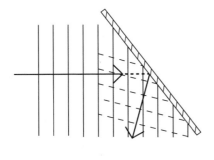

....

ii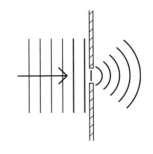

....

iii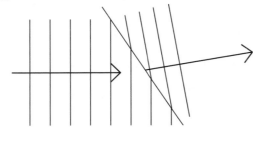

....

iv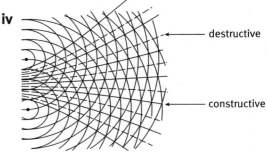

....

b The picture below right shows a water wave going from deep water to shallower water.

Choose words from this list to complete the sentences.

shallow	deep	frequency
wavelength	shorter	longer
slowed	reduced	

Water waves travel faster in .... water than in .... water. As the wave

crosses the boundary, the .... of the wave stays the same but the wavelength gets

.... This is because the waves are .... down; the progress of each wave is

.... by the time the next wave arrives.

c The wave in the picture is refracted.

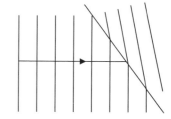

   i Put a tick (✓) in the region where the waves have a shorter wavelength.

   ii Are the waves slower or faster in this region? _____

   iii Label the two regions 'slow' and 'fast' to describe the speed of the waves.

   iv The arrow shows the direction of the wave before the boundary. Draw another arrow to show the direction of the wave after it has crossed the boundary.

   v The statements below explain why its direction changes. They are out of sequence. Use arrows to join the boxes in the correct order. The first one has been done for you.

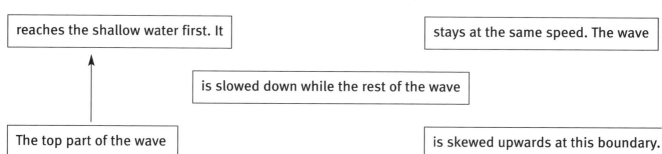

d Waves will spread out when they pass through a gap in a barrier.

wave approaching barrier   wave after it has passed through

   i What is the name of this effect?

   _____

   ii The picture on the right shows some waves approaching barriers and the waves after they have passed through the barrier.

   Draw lines to match each barrier with the wave after it has passed through.

   iii What happens to the spread of the wave as the gap gets wider?

   _____
   _____

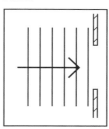 

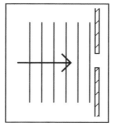 

e Complete these sentences. Draw a ring round the correct **bold** words.

Diffraction is most noticeable when the width of the slit is about the same size as the

wavelength / amplitude of the wave. Light has a very **long / short** wavelength. A light wave will diffract

when it passes through a very **narrow / wide** slit.

P6 The wave model of radiation – Higher

4 Sound and light – interference

a Interference is a wave effect.

→ Link up all the boxes below that describe **constructive interference**. Try to do them in a logical order. The first one has been done for you.

→ Now link up all the boxes that describe **destructive interference**. Try to do them in a logical order. Use a different colour.

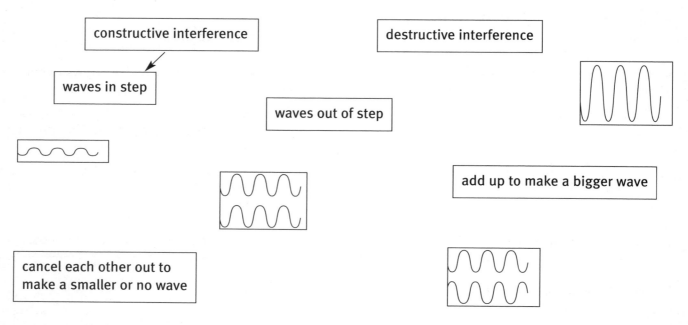

b The picture shows two loudspeakers connected to a signal generator. Each speaker is producing the same pure tone. The curved lines show a snapshot of the peaks of the sound waves.

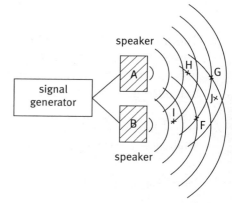

i Fraser is standing at the point marked F.
Complete these sentences. Draw a (ring) round the correct **bold** words.

When a peak from loudspeaker A reaches the point F,

a **peak / trough** from loudspeaker B reaches the same point.

The waves will **add up / cancel out**. Fraser will hear a sound that

is **louder / quieter** than one of the loudspeakers on its own.

ii There are people standing at the points G to J. Draw a circle round each person who is at a point of constructive interference.

iii Interference is a feature of waves. Does this experiment suggest that sound is a wave?

c We can also get interference effects with light.
Draw a (ring) round the piece of apparatus that is often used to produce an interference pattern.

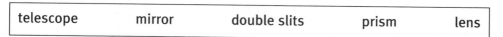

| telescope | mirror | double slits | prism | lens |

The wave model of radiation – Higher   P6

5 Is light a wave?

a Put a tick (✓) in the correct columns to show which behaviour is good evidence for sound and light being waves.

Behaviour of light	This happens to waves	Also happens to particles
reflected from a mirror		
diffracted through tiny gaps		
refracted as it passes from air to glass		
interference when two narrow slits produce bright and dark fringes		

Behaviour of sound	This happens to waves	Also happens to particles
produces echoes		
diffracted through an open door		
refracted between hot and cold air		
interference when two speakers produce loud and quiet regions		

b The picture shows a light ray at a boundary between air and water.

   i Light travels more slowly in water. Label the two sides of the boundary to show which side is air and which side is water.

   ii This ray of light is travelling from air to water. Add an arrow to the ray to show its direction.

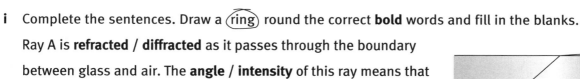

c The pictures below right show two rays of light (A and B) meeting the boundary between glass and air.

   i Complete the sentences. Draw a (ring) round the correct **bold** words and fill in the blanks.

   Ray A is **refracted / diffracted** as it passes through the boundary between glass and air. The **angle / intensity** of this ray means that it only just gets out of the glass. Ray B strikes the boundary at a **shallower / steeper** angle. It **can / cannot** get out of the glass. Instead, it is **reflected / absorbed** back into the glass. This is called

   T .... I .... R ....

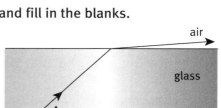

   ii Draw in the path of ray B after it has been reflected.

d The picture shows an optical fibre with a ray passing in at the left-hand end.

   i Draw the path of this ray as it goes through the fibre.
   ii Give two uses of optical fibres.

   → ....

   → ....

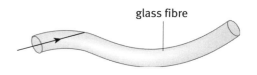

6 Electromagnetic waves

a Complete the sentences. Fill in the gaps and draw a ring round the correct bold words.

Light is part of a larger spectrum of **waves** / **sounds** called the E _____ S _____.
All of these waves travel at **the same** / **their own** speed in a vacuum. It is exactly **300 000** / **0.1** km/s. The shortest waves are **gamma rays** / **X-rays** whose wavelengths are around a billionth of a millimetre (about the size of a nucleus in **an atom** / **a plant cell**). The longest waves are **radio** / **ultraviolet** waves whose wavelengths are a few kilometres (the size of a small **town** / **person**).

b The picture shows the electromagnetic spectrum.

  i Fill in the boxes with the names of the regions. Visible light has been done for you.

  ii Draw lines to link each region with one or more of these uses.

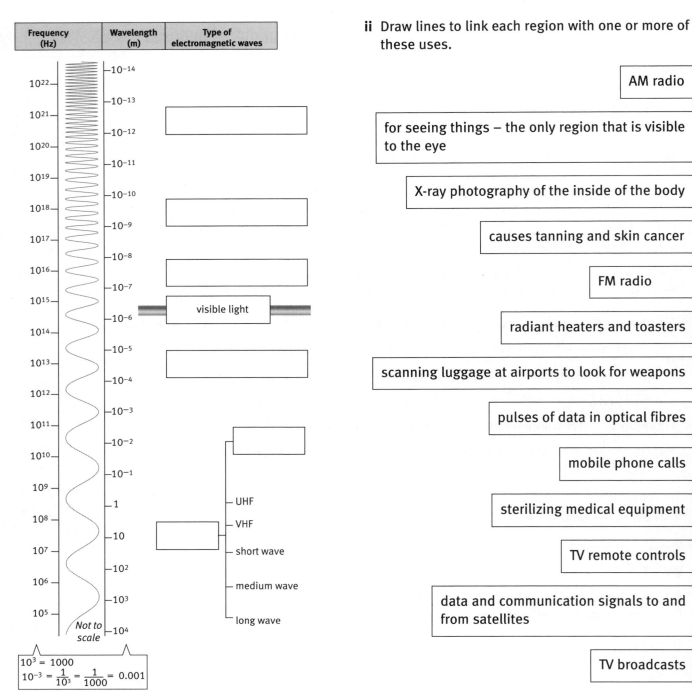

The wave model of radiation – Higher P6

7 The risky side of the rainbow

a There are three types of radiation with a wavelength that is shorter than visible light. The table represents this part of the electromagnetic spectrum.

Part of the electromagnetic spectrum
visible light

  i Fill in the names of the three types of radiation. Put the one with the shortest wavelength at the top.

  ii Is the frequency of this radiation higher or lower than the frequency of visible light? _____

  iii The radiation is carried as photons. Do these photons have more or less energy than the photons of visible light?

  iv Put a tick (✓) next to the part of the spectrum where a photon has the most energy.

b Choose words from this list to complete the sentences.

| cells | ionizing | high | low | electrons |
| molecules | randomizing | | ionize | photons |

When electromagnetic waves with a _____ frequency strike atoms, they can _____ the atoms. Their _____ carry enough energy to knock out _____. This can alter the material. If this happens in living _____, they can be damaged or destroyed. You should take precautions to reduce your exposure to these _____ radiations.

c Val puts on sunblock when she goes on the beach on hot days.

Complete these sentences. Draw a ring round the correct **bold** words.

The sunblock protects Val by **absorbing / transmitting** some of the ultraviolet radiation. Some radiation still reaches her skin. It has the same **wavelength / intensity** as before but a lower **wavelength / intensity**. So its photons have the **same / more** energy and the number getting through has been **reduced / unaffected**. This **reduces / increases** the risk of damage to her skin cells.

d A radiologist uses X-rays to look for broken bones. Explain why

  i the radiologist stands behind a protective screen _____

  ii the radiologist wears a badge that measures the radiation dose _____

P6 The wave model of radiation – Higher

8 Infrared

a The picture shows white light from a filament lamp passing through a prism.

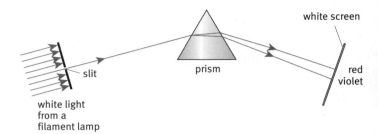

i Complete these sentences. Draw a ring round the correct **bold** words.

The light is **dispersed** / **combined** in the prism. This produces a **spectrum** / **reflection** of colours.

ii The speed of light in air is the same for all the colours. However, dispersion shows that the speed is different for different colours in glass. Which colour is the slowest in glass? _____

iii An infrared detector is put in front of the screen. Put an X on the diagram to show where it would detect radiation.

b i Now draw a line to match each type of object (left) with a description of the radiation it emits (middle). One has been done for you.

ii Now draw lines to link each description to two examples.

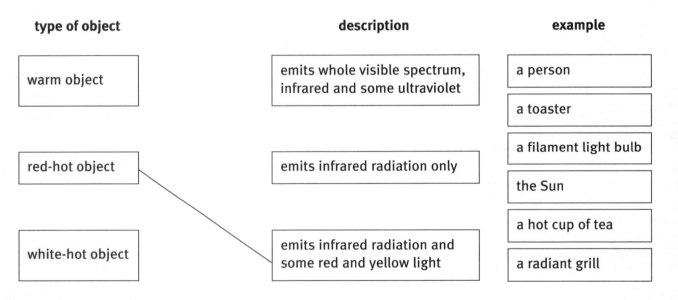

c Solve the clues to fill in the missing words about infrared (IR) radiation.

1 The type of radio wave that is next to infrared in the spectrum
2 Optical fibres allow infrared to pass through. They are . . .
3 This is lower for infrared than for visible light.
4 A type of imaging that can find warm bodies in the dark or under rubble
5 What infrared radiation will do to your skin when you absorb it
6 A _____ control; it uses an infrared beam to change TV channels
7 How the wavelength of infrared compares to visible light
8 The colour in the visible spectrum that is next to infrared

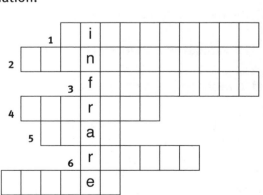

The wave model of radiation – Higher P6

9 Microwaves, radio waves, and SETI

a Look at the uses of microwaves below. Draw a line to match each use with a specific property of microwaves. The first one has been done for you.

carrying long distance telephone calls	They pass through the atmosphere.
cooking food	They can be beamed in a straight line.
carrying satellite TV signals	Some wavelengths are absorbed by water molecules.
carrying mobile phone signals	They can be produced by small electronic circuits.

b Look at the diagram. It shows the percentage of each type of radiation that reaches the ground from outside the Earth's atmosphere.

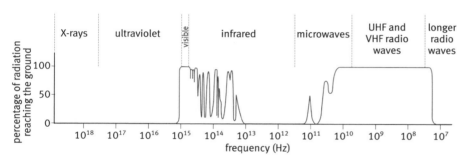

i Put an 'A' under the types of radiation that are absorbed by the atmosphere.

ii Put a 'T' under the frequencies that are transmitted by the atmosphere.

iii Suggest two types of scientist who make use of the microwave 'window' in the atmosphere.

→ _____   → _____

c Look at the statements on the left. Each one applies to either sound waves or electromagnetic waves. Draw a line to link each one with the type of wave it describes. The first one has been done for you.

| They are longitudinal waves. |
| They travel through a vacuum. |
| Their speed is about 300 000 km/s in air. | sound wave |
| Their speed is about 300 m/s in air. |
| They need some matter (solid, liquid, or gas) to carry them. | electromagnetic wave |
| They are transverse waves. |

P6 The wave model of radiation – Higher

10 Getting waves to carry information: AM

Waves can transmit information (a signal) if the amplitude of the wave is changed to match the signal.

a Draw a line to match each description with the correct wave.

audio frequency wave (AF)	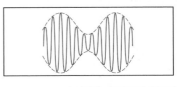
amplitude modulated wave (AM)	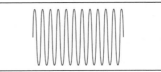
radio frequency carrier wave (RF)	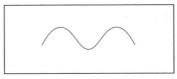

b Complete the sentences below. Draw a (ring) round the correct **bold** words.

Most humans can hear sounds up to a **frequency** / **wavelength** of about 20 kHz. A radio carrier wave has to have a frequency that is much **greater** / **smaller** than this. This means that there is a number of complete cycles of **carrier** / **audio** wave for each cycle of the **carrier** / **audio** wave.

c In the UK, there are dozens of radio stations that broadcast in the Medium Wave band, which covers frequencies from 620 kHz to 1630 kHz. Each radio station has its own broadcast frequency. Explain why

   i each station needs its own frequency: _____

   ii a radio receiver needs a tuning circuit: _____

d The list on the left shows some effects on AM broadcasts in the Long Wave band. The causes are shown on the right. Draw a line to link each effect to its cause.

The signal can pick up stray pops and buzzes from electrical equipment nearby.	The wave has a long wavelength which diffracts around geographical features.
The signal strength varies on motorways.	The wave is amplitude modulated so any interference that affects the amplitude becomes part of the signal.
The radio stations can be received behind hills and homes.	Waves from two transmitters can interfere with each other and produce peaks and troughs in the signal strength.

c The diagram shows a simple radio transmitter (on the left) and receiver (on the right).

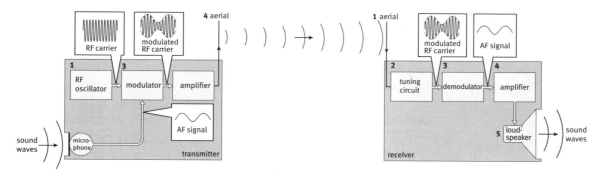

Note: you do not need to remember the details of these diagrams. They are to help you understand how the modulated wave is made and what it looks like.

i The boxes below describe the stages in producing and broadcasting an AM signal. They are out of sequence. Draw arrows to join the boxes in the correct sequence. The first one has been done for you.

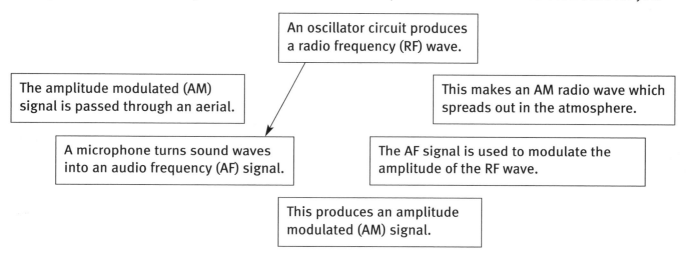

ii The radio receiver picks up the AM radio wave and produces a sound wave. The list on the left shows some of the parts of a radio receiver. They are described on the right. Draw a line to match each part with its description. The first one has been done for you.

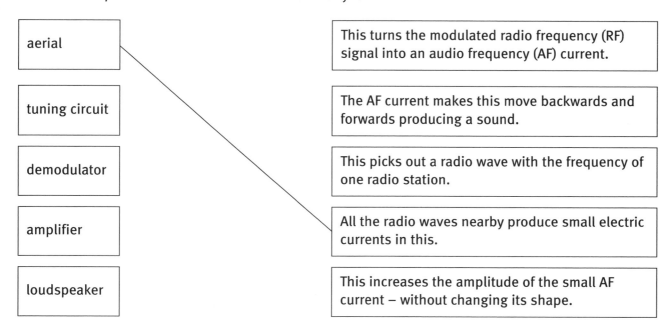

P6 The wave model of radiation – Higher

11 FM and digital

a Use these words to complete the descriptions below. (You need to use some words more than once.)

| audio frequency (AF) | frequency | amplitude modulated (AM) |
| amplitude | frequency modulated (FM) | |

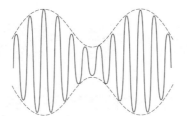

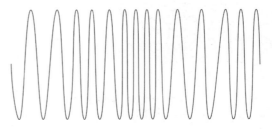

This signal is _____

The changes in the _____

exactly match the changes in the original

_____ signal.

This signal is _____

The changes in the _____

exactly match the changes of the original

_____ signal.

b Both the signals in the previous question are analogue signals. They change continuously. A digital signal does not vary continuously. The voltage is sampled many times a second and each value is given a code. The codes are **binary**. All the numbers contain only two digits – 0 (no pulse) and 1 (a pulse).

The diagram shows how an analogue signal can be sampled to give a digital signal.

Follow these steps to complete the diagram.

In the boxes:

- write in the value of the voltage – to the nearest whole number
- write in the binary number that has the value in the box above
- draw the pulses to represent the binary number

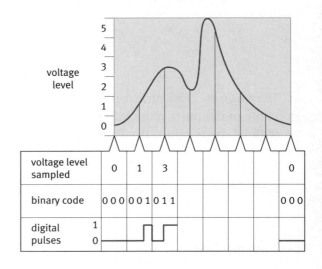

c complete the sentences below. Draw a ring round the correct bold words.

Digital signals can be sent along optical **strings / fibres**. The 1s and 0s are used to switch a **laser / lamp** on and off; it **emits / absorbs** millions of **infrared / ultraviolet** pulses every second. Fibres can be many **kilometres / millimetres** long. They are made from special **glass / steel** that is extremely pure so that hardly any of the radiation is **absorbed / transmitted**.

12 Going digital

a Digital signals are replacing many analogue signals. But analogue signals are still used.

The table shows some technologies which have analogue and digital versions.

⇨ Fill in the **Digital** and **Analogue** columns to show the media and methods used in each version.

Use the words and abbreviations in this list. Each word is used once only.

| DAB | 35 mm film | Hard disk/MP3 | AM | Cassette | CCD |

Technology	Digital	Analogue	Advantages of digital
cameras			
talk radio			
portable music player			

⇨ For each technology, write in an advantage of using digital signals.

b The diagram below shows the transmission of an analogue signal and a digital signal.

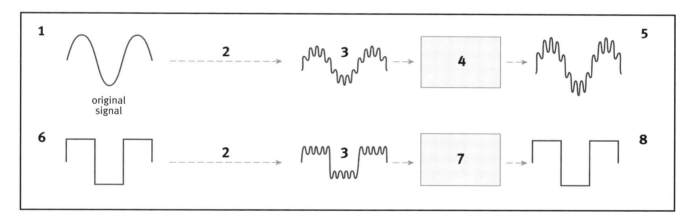

i Match these labels with the numbers on the diagram. Write the correct number in each box.

☐ weak signal with noise   ☐ regenerated signal   ☐ amplifier   ☐ digital signal

☐ amplified signal with noise   ☐ analogue signal   ☐ transmission   ☐ regenerator

ii Which word can be defined as 'low level, unwanted waves that contaminate the original signal'?

N..........

iii Which type of signal can produce a perfect copy of itself at the destination?

iv Explain why this is such an advantage. ....

This page is blank